Astronomy of Today

by Cecil G. Dolmage

The object of this book is to give an account of the science of Astronomy, as it is known at the present day, in a manner acceptable to the general reader.

It is too often supposed that it is impossible to acquire any useful knowledge of Astronomy without much laborious study, and without adventuring into quite a new world of thought. The reasoning applied to the study of the celestial orbs is, however, of no different order from that which is employed in the affairs of everyday life. The science of mathematics is perhaps responsible for the idea that some kind of difference does exist; but mathematical processes are, in effect, no more than ordinary logic in concentrated form, the shorthand of reasoning, so to speak. I have attempted in the following pages to take the main facts and theories of Astronomy out of those mathematical forms which repel the general reader, and to present them in the ordinary language of our workaday world.

The few diagrams introduced are altogether supplementary, and are not connected with the text by any wearying cross-references. Each diagram is complete in itself, being intended to serve as a pictorial aid, in case the wording of the text should not have perfectly conveyed the desired meaning. The full page illustrations are also described as adequately as possible at the foot of each.

As to the coloured frontispiece, this must be placed in a category by itself. It is the work of the artist as distinct from the scientist.

The book itself contains incidentally a good deal of matter concerned with the Astronomy of the past, the introduction of which has been found necessary in order to make clearer the Astronomy of our time.

It would be quite impossible for me to enumerate here the many sources from which information has been drawn. But I acknowledge my especial indebtedness to Professor F.R. Moulton's Introduction to Astronomy (Macmillan, 1906), to the works on Eclipses of the late Rev. S.J. Johnson and of Mr. W.T. Lynn, and to the excellent Journals of the British Astronomical Association. Further, for those grand questions concerned with the Stellar Universe at large, I owe a very deep debt to the writings of the famous

American astronomer, Professor Simon Newcomb, and of our own countryman, Mr. John Ellard Gore; to the latter of whom I am under an additional obligation for much valuable information privately rendered.

In my search for suitable illustrations, I have been greatly aided by the kindly advice of Mr. W. H. Wesley, the Assistant Secretary of the Royal Astronomical Society. To those who have been so good as to permit me to reproduce pictures and photographs, I desire to record my best thanks as follows:--To the French Artist, Mdlle. Andr 閑 Moch; to the Astronomer Royal; to Sir David Gill, K.C.B., LL.D., F.R.S.; to the Council of the Royal Astronomical Society; to Professor E.B. Frost, Director of the Yerkes Observatory; to M.P. Puiseux, of the Paris Observatory; to Dr. Max Wolf, of Heidelberg; to Professor Percival Lowell; to the Rev. Theodore E.R. Phillips, M.A., F.R.A.S.; to Mr. W.H. Wesley; to the Warner and Swasey Co., of Cleveland, Ohio, U.S.A.; to the publishers of Knowledge, and to Messrs. Sampson, Low & Co. For permission to reproduce the beautiful photograph of the Spiral Nebula in Canes Venatici (Plate XXII.), I am indebted to the distinguished astronomer, the late Dr. W.E. Wilson, D.Sc., F.R.S., whose untimely death, I regret to state, occurred in the early part of this year.

Finally, my best thanks are due to Mr. John Ellard Gore, F.R.A.S., M.R.I.A., to Mr. W.H. Wesley, and to Mr. John Butler Burke, M.A., of Cambridge, for their kindness in reading the proof-sheets.

CECIL G. DOLMAGE.

LONDON, S.W., August 4, 1908.

CONTENTS

CHAPTER I

THE ANCIENT VIEW

It is never safe, as we know, to judge by appearances, and this is perhaps more true of astronomy than of anything else.

For instance, the idea which one would most naturally form of the earth and heaven is that the solid earth on which we live and move extends to a great distance in every direction, and that the heaven is an immense dome upon the inner surface of which the stars are fixed. Such must needs have been the idea of the universe held by men in the earliest times. In their view the earth was of paramount importance. The sun and moon were mere lamps for the day and for the night; and these, if not gods themselves, were at any rate under the charge of special deities, whose task it was to guide their motions across the vaulted sky.

Little by little, however, this simple estimate of nature began to be overturned. Difficult problems agitated the human mind. On what, for instance, did the solid earth rest, and what prevented the vaulted heaven from falling in upon men and crushing them out of existence? Fantastic myths sprang from the vain attempts to solve these riddles. The Hindoos, for example, imagined the earth as supported by four elephants which stood upon the back of a gigantic tortoise, which, in its turn, floated on the surface of an elemental ocean. The early Western civilisations conceived the fable of the Titan Atlas, who, as a punishment for revolt against the Olympian gods, was condemned to hold up the expanse of sky for ever and ever.

Later on glimmerings of the true light began to break in upon men. The Greek philosophers, who busied themselves much with such matters, gradually became convinced that the earth was spherical in shape, that is to say, round like a ball. In this opinion we now know that they were right; but in their other important belief, viz. that the earth was placed at the centre of all things, they were indeed very far from the truth.

By the second century of the Christian era, the ideas of the early philosophers had become hardened into a definite theory, which, though it appears very incorrect to us to-day, nevertheless demands exceptional notice from the fact that it was everywhere accepted as the true explanation until so late as some four centuries ago. This theory of the universe is known by the name of the Ptolemaic System, because it was first set forth in definite terms by one of the most famous of the astronomers of antiquity, Claudius Ptolemys Pelusinensis (100-170 A.D.), better known as Ptolemy of Alexandria.

In his system the Earth occupied the centre; while around it circled in order outwards the Moon, the planets Mercury and Venus, the Sun, and then the planets Mars, Jupiter, and Saturn. Beyond these again revolved the background of the heaven, upon which it was believed that the stars were fixed--

"Stellis ardentibus aptum,"

as Virgil puts it (see Fig. 1).

The Ptolemaic system persisted unshaken for about fourteen hundred years after the death of its author. Clearly men were flattered by the notion that their earth was the most important body in nature, that it stood still at the centre of the universe, and was the pivot upon which all things revolved.

CHAPTER II

THE MODERN VIEW

It is still well under four hundred years since the modern, or Copernican, theory of the universe supplanted the Ptolemaic, which had held sway during so many centuries. In this new theory, propounded towards the middle of the sixteenth century by Nicholas Copernicus (1473-1543), a Prussian astronomer, the earth was dethroned from its central position and considered merely as one of a number of planetary bodies which revolve around the sun. As it is not a part of our purpose to follow in detail the history of the science, it seems advisable to begin by stating in a broad fashion the conception of the universe as accepted and believed in to-day.

The Sun, the most important of the celestial bodies so far as we are concerned, occupies the central position; not, however, in the whole universe, but only in that limited portion which is known as the Solar System. Around it, in the following order outwards, circle the planets Mercury, Venus, the Earth, Mars, Jupiter, Saturn, Uranus, and Neptune (see Fig. 2, p. 21). At an immense distance beyond the solar system, and scattered irregularly through the depth of space, lie the stars. The two first-mentioned members of the solar system, Mercury and Venus, are known as the Inferior Planets; and in their courses about the sun, they always keep well inside the path along which our earth moves. The remaining members (exclusive of the earth) are called Superior Planets, and their paths lie all outside that of the earth.

The five planets, Mercury, Venus, Mars, Jupiter, and Saturn, have been known from all antiquity. Nothing then can bring home to us more strongly the immense advance which has taken place in astronomy during modern times than the fact that it is only 127 years since observation of the skies first added a planet to that time-honoured number. It was indeed on the 13th of March 1781, while engaged in observing the constellation of the Twins, that the justly famous Sir William Herschel caught sight of an object which he did not recognise as having met with before. He at first took it for a comet, but observations of its movements during a few days showed it to be a planet. This body, which the power of the telescope alone had thus shown to belong to the solar family, has since become known to science under the name of Uranus. By its discovery the hitherto accepted limits of the solar system were at once pushed out to twice their former extent, and the hope naturally arose that other planets would quickly reveal themselves in the immensities beyond.

For a number of years prior to Herschel's great discovery, it had been noticed that the distances at which the then known planets circulated appeared to be arranged in a somewhat orderly progression outwards from the sun. This seeming plan, known to astronomers by the name of Bode's Law, was closely confirmed by the distance of the new planet Uranus. There still lay, however, a broad gap between the planets Mars and Jupiter. Had another planet indeed circuited there, the solar system would have presented an appearance of almost perfect order. But the void between Mars and Jupiter was unfilled; the space in which one would reasonably expect to find another planet circling was unaccountably empty.

On the first day of the nineteenth century the mystery was however explained, a body being discovered[1] which revolved in the space that had hitherto been considered planetless. But it was a tiny globe hardly worthy of the name of planet. In the following year a second body was discovered revolving in the same space; but it was even smaller in size than the first. During the ensuing five years two more of these little planets were discovered. Then came a pause, no more such bodies being added to the system until half-way through the century, when suddenly the discovery of these so-called "minor planets" began anew. Since then additions to this portion of our system have rained thick and fast. The small bodies have received the name of Asteroids or Planetoids; and up to the present time some six hundred of them are known to exist, all revolving in the previously unfilled space between Mars and Jupiter.

In the year 1846 the outer boundary of the solar system was again extended by the discovery that a great planet circulated beyond Uranus. The new body, which received the name of Neptune, was brought to light as the result of calculations made at the same time, though quite independently, by the Cambridge mathematician Adams, and the French astronomer Le Verrier. The discovery of Neptune differed, however, from that of Uranus in the following respect. Uranus was found merely in the course of ordinary telescopic survey of the heavens. The position of Neptune, on the other hand, was predicted as the result of rigorous mathematical investigations undertaken with the object of fixing the position of an unseen and still more distant body, the attraction of which, in passing by, was disturbing the position of Uranus in its revolution around the sun. Adams actually completed his investigation first; but a delay at Cambridge in examining that portion of the sky, where he announced that the body ought just then to be, allowed France to snatch the honour of discovery, and the new planet was found by the observer Galle at Berlin, very near the place in the heavens which Le Verrier had mathematically predicted for it.

Nearly fifty years later, that is to say, in our own time, another important planetary discovery was made. One of the recent additions to the numerous and constantly increasing family of the asteroids, a tiny body brought to light in 1898, turned out after all not to be circulating in the customary space between Mars and Jupiter, but actually in that between our earth and Mars. This body is very small, not more than about twenty miles across. It has received the name of Eros (the Greek equivalent for Cupid), in allusion to its insignificant size as compared with the other leading members of the system.

This completes the list of the planets which, so far, have revealed themselves to us. Whether others exist time alone will show. Two or three have been suspected to revolve beyond the path of Neptune; and it has even been asserted, on more than one occasion, that a planet circulates nearer to the sun than Mercury. This supposed body, to which the name of "Vulcan" was provisionally given, is said to have been "discovered" in 1859 by a French doctor named Lescarbault, of Orleans; but up to the present there has been no sufficient evidence of its existence. The reason why such uncertainty should exist upon this point is easy enough to understand, when we consider the overpowering glare which fills our atmosphere all around the sun's place in the sky. Mercury, the nearest known planet to the sun, is for this reason always very difficult to see; and even when, in its course, it gets sufficiently far from the sun to be left for a short time above the horizon after sunset, it is by no means an easy object to observe on account of the mists which usually hang about low down near the earth. One opportunity, however, offers itself from time to time to solve the riddle of an "intra-Mercurial" planet, that is to say, of a planet which circulates within the path followed by Mercury. The opportunity in question is furnished by a total eclipse of the sun; for when, during an eclipse of that kind, the body of the moon for a few minutes entirely hides the sun's face, and the dazzling glare is thus completely cut off, astronomers are enabled to give an unimpeded, though all too hurried, search to the region close around. A goodly number of total eclipses of the sun have, however, come and gone since the days of Lescarbault, and no planet, so far, has revealed itself in the intra-Mercurial space. It seems, therefore, quite safe to affirm that no globe of sufficient size to be seen by means of our modern telescopes circulates nearer to the sun than the planet Mercury.

Next in importance to the planets, as permanent members of the solar system, come the relatively small and secondary bodies known by the name of Satellites. The name satellite is derived from a Latin word signifying an attendant; for the bodies so-called move along always in close proximity to their respective "primaries," as the planets which they accompany are technically termed.

The satellites cannot be considered as allotted with any particular regularity among the various members of the system; several of the planets, for instance, having a goodly number of these bodies accompanying them, while others have but one or two, and some again have none at all. Taking the planets in their order of distance outward from the Sun, we find that neither Mercury nor Venus are provided with satellites; the Earth has only one, viz. our neighbour the Moon; while Mars has but two tiny ones, so small indeed that one might imagine them to be merely asteroids, which had wandered out of their proper region and attached themselves to that planet. For the rest, so far as we at present know, Jupiter possesses seven,[2] Saturn ten, Uranus four, and Neptune one. It is indeed possible, nay more, it is extremely probable, that the two last-named planets have a greater number of these secondary bodies revolving around them; but, unfortunately, the Uranian and Neptunian systems are at such immense distances from us, that even the magnificent telescopes of to-day can extract very little information concerning them.

From the distribution of the satellites, the reader will notice that the planets relatively near to the sun are provided with few or none, while the more distant planets are richly endowed. The conclusion, therefore, seems to be that nearness to the sun is in some way unfavourable either to the production, or to the continued existence, of satellites.

A planet and its satellites form a repetition of the solar system on a tiny scale. Just as the planets revolve around the sun, so do these secondary bodies revolve around their primaries. When Galileo, in 1610, turned his newly invented telescope upon Jupiter, he quickly recognised in the four circling moons which met his gaze, a miniature edition of the solar system.

Besides the planets and their satellites, there are two other classes of bodies which claim membership of the solar system. These are Comets and Meteors. Comets differ from the bodies which we have just been describing in that they appear filmy and transparent, whereas the others are solid and opaque. Again, the paths of the planets around the sun and of the satellites around their primaries are not actually circles; they are ovals, but their ovalness is not of a marked degree. The paths of comets on the other hand are usually very oval; so that in their courses many of them pass out as far as the known limits of the solar system, and even far beyond. It should be mentioned that nowadays the tendency is to consider comets as permanent members of the system, though this was formerly not by any means an article of faith with astronomers.

Meteors are very small bodies, as a rule perhaps no larger than pebbles, which move about unseen in space, and of which we do not become aware until they arrive very close to the earth. They are then made visible to us for a moment or two in consequence of being heated to a white heat by the friction of rushing through the atmosphere, and are thus usually turned into ashes and vapour long before they reach the surface of our globe. Though occasionally a meteoric body survives the fiery ordeal, and reaches the earth more or less in a solid state to bury itself deep in the soil, the majority of these celestial visitants constitute no source of danger whatever for us. Any one who will take the trouble to gaze at the sky for a short time on a clear night, is fairly certain to be rewarded with the view of a meteor. The impression received is as if one of the stars had suddenly left its accustomed place, and dashed across the heavens, leaving in its course a trail of light. It is for this reason that meteors are popularly known under the name of "shooting stars."

[1] By the Italian astronomer, Piazzi, at Palermo.

[2] Probably eight. (See note, page 232.)

CHAPTER III

THE SOLAR SYSTEM

We have seen, in the course of the last chapter, that the solar system is composed as follows:--there is a central body, the sun, around which revolve along stated paths a number of important bodies known as planets. Certain of these planets, in their courses, carry along in company still smaller bodies called satellites, which revolve around them. With regard, however, to the remaining members of the system, viz. the comets and the meteors, it is not advisable at this stage to add more to what has been said in the preceding chapter. For the time being, therefore, we will devote our attention merely to the sun, the planets, and the satellites.

Of what shape then are these bodies? Of one shape, and that one alone which appears to characterise all solid objects in the celestial spaces: they are spherical, which means round like a ball.

Each of these spherical bodies rotates; that is to say, turns round and round, as a top does when it is spinning. This rotation is said to take place "upon an axis," a statement which may be explained as follows:--Imagine a ball with a knitting-needle run right through its centre. Then imagine this needle held pointing in one fixed direction while the ball is turned round and round. Well, it is the same thing with the earth. As it journeys about the sun, it keeps turning round and round continually as if pivoted upon a mighty knitting needle transfixing it from North Pole to South Pole. In reality, however, there is no such material axis to regulate the constant direction of the rotation, just as there are no actual supports to uphold the earth itself in space. The causes which keep the celestial spheres poised, and which control their motions, are far more wonderful than any mechanical device.

At this juncture it will be well to emphasise the sharp distinction between the terms rotation and revolution. The term "rotation" is invariably used by astronomers to signify the motion which a celestial body has upon an axis; the term "revolution," on the other hand, is used for the movement of one celestial body around another. Speaking of the earth, for instance, we say, that it rotates on its axis, and that it revolves around the sun.

So far, nothing has been said about the sizes of the members of our system. Is there any stock size, any pattern according to which they may be judged? None whatever! They vary enormously. Very much the largest of all is the Sun, which is several hundred times larger than all the planets and satellites of the system rolled together. Next comes Jupiter and afterwards the other planets in the following order of size:--Saturn, Uranus, Neptune, the Earth, Venus, Mars, and Mercury. Very much smaller than any of these are the asteroids, of which Ceres, the largest, is less than 500 miles in diameter. It is, by the way, a strange fact that the zone of asteroids should mark the separation of the small planets from the giant ones. The following table, giving roughly the various diameters of the sun and the principal planets in miles, will clearly illustrate the great discrepancy in size which prevails in the system.

Sun 866,540 miles Mercury 2,765 " Venus 7,826 " Earth 7,918 " Mars 4,332 "

Jupiter 87,380 " Saturn 73,125 " Uranus[3] 34,900 " Neptune[3] 32,900 "

It does not seem possible to arrive at any generalisation from the above data, except it be to state that there is a continuous increase in size from Mercury to the earth, and a similar decrease in size from Jupiter outwards. Were Mars greater than the earth, the planets could then with truth be said to increase in size up to Jupiter, and then to decrease. But the zone of asteroids, and the relative smallness of Mars, negative any attempt to regard the dimensions of the planets as an orderly sequence.

Next with respect to relative distance from the sun, Venus circulates nearly twice as far from it as Mercury, the earth nearly three times as far, and Mars nearly four times. After this, just as we found a sudden increase in size, so do we meet with a sudden increase in distance. Jupiter, for instance, is about thirteen times as far as Mercury, Saturn about twenty-five times, Uranus about forty-nine times, and Neptune about seventy-seven. (See Fig. 2, p. 21.)

It will thus be seen how enormously the solar system was enlarged in extent by the discovery of the outermost planets. The finding of Uranus plainly doubled its breadth; the finding of Neptune made it more than half as broad again. Nothing indeed can better show the import of these great discoveries than to take a pair of compasses and roughly set out the above relative paths in a series of concentric circles upon a large sheet of paper, and then to consider that the path of Saturn was the supposed boundary of our solar system prior to the year 1781.

We have seen that the usual shape of celestial bodies themselves is spherical. Of what form then are their paths, or orbits, as these are called? One might be inclined at a venture to answer "circular," but this is not the case. The orbits of the planets cannot be regarded as true circles. They are ovals, or, to speak in technical language, "ellipses." Their ovalness or "ellipticity" is, however, in each case not by any means of the same degree. Some orbits--for instance, that of the earth--differ only slightly from circles; while others--those of Mars or Mercury, for example--are markedly elliptic. The orbit of the tiny planet Eros is, however, far and away the most elliptic of all, as we shall see when we come to deal with that little planet in detail.

It has been stated that the sun and planets are always rotating. The various rates at which they do so will, however, be best appreciated by a comparison with the rate at which the earth itself rotates.

But first of all, let us see what ground we have, if any, for asserting that the earth rotates at all?

If we carefully watch the heavens we notice that the background of the sky, with all the brilliant objects which sparkle in it, appears to turn once round us in about twenty-four hours; and that the pivot upon which this movement takes place is situated somewhere near what is known to us as the Pole Star. This was one of the earliest facts noted with regard to the sky; and to the men of old it therefore seems as if the heavens and all therein were always revolving around the earth. It was natural enough for them to take this view, for they had not the slightest idea of the immense distance of the celestial bodies, and in the absence of any knowledge of the kind they were inclined to imagine them comparatively near. It was indeed only after the lapse of many centuries, when men had at last realised the enormous gulf which separated them from even the nearest object in the sky, that a more reasonable opinion began to prevail. It was then seen that this revolution of the heavens about the earth could be more easily and more satisfactorily explained by supposing a mere rotation of the solid earth about a fixed axis, pointed in the direction of the polar star. The probability of such a rotation on the part of the earth itself was further strengthened by the observations made with the telescope. When the surfaces of the sun and planets were carefully studied these bodies were seen to be rotating. This being the case, there could not surely be much hesitation in granting that the earth rotated also; particularly when it so simply explained the daily movement of the sky, and saved men from the almost inconceivable notion that the whole stupendous vaulted heaven was turning about them once in every twenty-four hours.

If the sun be regularly observed through a telescope, it will gradually be gathered from the slow displacement of sunspots across its face, their disappearance at one edge and their reappearance again at the other edge, that it is rotating on an axis in a period of about twenty-six days. The movement, too, of various well-known markings on the surfaces of the planets Mars, Jupiter, and Saturn proves to us that these bodies are rotating in periods, which are about twenty-four hours for the first, and about ten hours for each of the other two. With regard, however, to Uranus and Neptune there is much more uncertainty, as these planets are at such great distances that even our best telescopes give but a confused view of the markings which they display; still a period of rotation of from ten to twelve hours appears to be accepted for them. On the other hand the constant blaze of sunlight in the neighbourhood of Mercury and Venus equally hampers astronomers in this quest. The older telescopic observers considered that the rotation periods of these two planets were about the same as that of the earth; but of recent years the opinion has been gaining ground that they turn round on their axes in exactly the same time as they revolve about the sun. This question is, however, a very doubtful one, and will be again referred to later on; but, putting it on one side, it will be seen from what we have said above, that the rotation periods of the other planets of our system are usually about twenty-four hours, or under. The fact that the rotation period of the sun should run into days need not seem extraordinary when one considers its enormous size.

The periods taken by the various planets to revolve around the sun is the next point which has to be considered. Here, too, it is well to start with the earth's period of revolution as the standard, and to see how the periods taken by the other planets compare with it.

The earth takes about 365-1/4 days to revolve around the sun. This period of time is known to us as a "year." The following table shows in days and years the periods taken by each of the other planets to make a complete revolution round the sun:--

Mercury about 88 days. Venus " 226 " Mars " 1 year and 321 days. Jupiter " 11 years and 313 days. Saturn " 29 years and 167 days. Uranus " 84 years and 7 days. Neptune " 164 years and 284 days.

From these periods we gather an important fact, namely, that the nearer a planet is to the sun the faster it revolves.

Compared with one of our years what a long time does an Uranian, or Neptunian, "year" seem? For instance, if a "year" had commenced in

Neptune about the middle of the reign of George II., that "year" would be only just coming to a close; for the planet is but now arriving back to the position, with regard to the sun, which it then occupied. Uranus, too, has only completed a little more than 1-1/2 of its "years" since Herschel discovered it.

Having accepted the fact that the planets are revolving around the sun, the next point to be inquired into is:--What are the positions of their orbits, or paths, relatively to each other?

Suppose, for instance, the various planetary orbits to be represented by a set of hoops of different sizes, placed one within the other, and the sun by a small ball in the middle of the whole; in what positions will these hoops have to be arranged so as to imitate exactly the true condition of things?

First of all let us suppose the entire arrangement, ball and hoops, to be on one level, so to speak. This may be easily compassed by imagining the hoops as floating, one surrounding the other, with the ball in the middle of all, upon the surface of still water. Such a set of objects would be described in astronomical parlance as being in the same plane. Suppose, on the other hand, that some of these floating hoops are tilted with regard to the others, so that one half of a hoop rises out of the water and the other half consequently sinks beneath the surface. This indeed is the actual case with regard to the planetary orbits. They do not by any means lie all exactly in the same plane. Each one of them is tilted, or inclined, a little with respect to the plane of the earth's orbit, which astronomers, for convenience, regard as the level of the solar system. This tilting, or "inclination," is, in the larger planets, greatest for the orbit of Mercury, least for that of Uranus. Mercury's orbit is inclined to that of the earth at an angle of about 7° that of Venus at a little over 3° that of Saturn 2-1/2° while in those of Mars, Neptune, and Jupiter the inclination is less than 2° But greater than any of these is the inclination of the orbit of the tiny planet Eros, viz. nearly 11°

The systems of satellites revolving around their respective planets being, as we have already pointed out, mere miniature editions of the solar system, the considerations so far detailed, which regulate the behaviour of the planets in their relations to the sun, will of necessity apply to the satellites very closely. In one respect, however, a system of satellites differs materially from a system of planets. The central body around which planets are in motion is self-luminous, whereas the planetary body around which a satellite revolves is not. True, planets shine, and shine very brightly too; as, for instance, Venus and Jupiter. But they do not give forth any light of their own, as the sun does; they merely reflect the sunlight which they receive from him. Putting this one fact aside, the analogy between the planetary system and a satellite system is remarkable. The satellites are spherical in form, and differ markedly in size; they rotate, so far as we know, upon their axes in varying times; they revolve around their governing planets in orbits, not circular, but elliptic; and these orbits, furthermore, do not of necessity lie in the same plane. Last of all the satellites revolve around their primaries at rates which are directly comparable with those at which the planets revolve around the sun, the rule in fact holding good that the nearer a satellite is to its primary the faster it revolves.

[3] As there seems to be much difference of opinion concerning the diameters of Uranus and Neptune, it should here be mentioned that the above figures are taken from Professor F.R. Moulton's Introduction to Astronomy (1906). They are there stated to be given on the authority of "Barnard's many measures at the Lick Observatory."

CHAPTER IV

CELESTIAL MECHANISM

As soon as we begin to inquire closely into the actual condition of the various members of the solar system we are struck with a certain distinction. We find that there are two quite different points of view from which these bodies can be regarded. For instance, we may make our estimates of them either as regards volume--that is to say, the mere room which they take up; or as regards mass--that is to say, the amount of matter which they contain.

Let us imagine two globes of equal volume; in other words, which take up an equal amount of space. One of these globes, however, may be composed of material much more tightly put together than in the other; or of greater density, as the term goes. That globe is said to be the greater of the two in mass. Were such a pair of globes to be weighed in scales, one globe in each pan, we should see at once, by its weighing down the other, which of the two was composed of the more tightly packed materials; and we should, in astronomical parlance, say of this one that it had the greater mass.

Volume being merely another word for size, the order of the members of the solar system, with regard to their volumes, will be as follows, beginning with the greatest:--the Sun, Jupiter, Saturn, Uranus, Neptune, the Earth, Venus, Mars, and Mercury.

With regard to mass the same order strangely enough holds good. The actual densities of the bodies in question are, however, very different. The densest or closest packed body of all is the Earth, which is about five and a half times as dense as if it were composed entirely of water. Venus follows next, then Mars, and then Mercury. The remaining bodies, on the other hand, are relatively loose in structure. Saturn is the least dense of all, less so than water. The density of the Sun is a little greater than that of water.

This method of estimating is, however, subject to a qualification. It must be remembered that in speaking of the Sun, for instance, as being only a little denser than water, we are merely treating the question from the point of view of an average. Certain parts of it in fact will be ever so much denser than water: those are the parts in the centre. Other portions, for instance, the outside portions, will be very much less dense. It will easily be understood that in all such bodies the densest or most compressed portions are to be found towards the centre; while the portions towards the exterior being less pressed upon, will be less dense.

We now reach a very important point, the question of Gravitation. Gravitation, or gravity, as it is often called, is the attractive force which, for instance, causes objects to fall to the earth. Now it seems rather strange that one should say that it is owing to a certain force that things fall towards the earth. All things seem to us to fall so of their own accord, as if it were quite natural, or rather most unnatural if they did not. Why then require a "force" to make them fall?

The story goes that the great Sir Isaac Newton was set a-thinking on this subject by seeing an apple fall from a tree to the earth. He then carried the train of thought further; and, by studying the movements of the moon, he reached the conclusion that a body even so far off as our satellite would be drawn towards the earth in the same manner. This being the case, one will naturally ask why the moon herself does not fall in upon the earth. The answer is indeed found to be that the moon is travelling round and round the earth at a certain rapid pace, and it is this very same rapid pace which keeps her from falling in upon us. Any one can test this simple fact for himself. If we tie a stone to the end of a string, and keep whirling it round and round fast enough, there will be a strong pull from the stone in an outward direction, and the string will remain tight all the time that the stone is being whirled. If, however, we gradually slacken the speed at which we are making the stone whirl, a moment will come at length when the string will become limp, and the stone will fall back towards our hand.

It seems, therefore, that there are two causes which maintain the stone at a regular distance all the time it is being steadily whirled. One of these is the continual pull inward towards our hand by means of the string. The other is the continual pull away from us caused by the rate at which the stone is travelling. When the rate of whirling is so regulated that these pulls exactly balance each other, the stone travels comfortably round and round, and shows no tendency either to fall back upon our hand or to break the string and fly away into the air. It is indeed precisely similar with regard to the moon. The continual pull of the earth's gravitation takes the place of the string. If the moon were to go round and round slower than it does, it would tend to fall in towards the earth; if, on the other hand, it were to go faster, it would tend to rush away into space.

The same kind of pull which the earth exerts upon the objects at its surface, or upon its satellite, the moon, exists through space so far as we know. Every particle of matter in the universe is found in fact to attract every other particle. The moon, for instance, attracts the earth also, but the controlling force is on the side of the much greater mass of the earth. This force of gravity or attraction of gravitation, as it is also called, is perfectly regular in its action. Its power depends first of all exactly upon the mass of the body which exerts it. The gravitational pull of the sun, for instance, reaches out to an enormous distance, controlling perhaps, in their courses, unseen planets circling far beyond the orbit of Neptune. Again, the strength with which the force of gravity acts depends upon distance in a regularly diminishing proportion. Thus, the nearer an object is to the earth, for instance, the stronger is the gravitational pull which it gets from it; the farther off it is, the weaker is this pull. If then the moon were to be brought nearer to the earth, the gravitational pull of the latter would become so much stronger that the moon's rate of motion would have also to increase in due proportion to prevent her from being drawn into the earth. Last of all, the point in a body from which the attraction of gravitation acts, is not necessarily the centre of the body, but rather what is known as its centre of gravity, that is to say, the balancing point of all the matter which the body contains.

It should here be noted that the moon does not actually revolve around the centre of gravity of the earth. What really happens is that both orbs revolve around their common centre of gravity, which is a point within the body of the earth, and situated about three thousand miles from its centre. In the same manner the planets and the sun revolve around the centre of gravity of the solar system, which is a point within the body of the sun.

The neatly poised movements of the planets around the sun, and of the satellites around their respective planets, will therefore be readily understood to result from a nice balance between gravitation and speed of motion.

The mass of the earth is ascertained to be about eighty times that of the moon. Our knowledge of the mass of a planet is learned from comparing the revolutions of its satellite or satellites around it, with those of the moon around the earth. We are thus enabled to deduce what the mass of such a planet would be compared to the earth's mass; that is to say, a study, for instance, of Jupiter's satellite system shows that Jupiter must have a mass nearly three hundred and eighteen times that of our earth. In the same manner we can argue out the mass of the sun from the movements of the planets and other bodies of the system around it. With regard, however, to Venus and Mercury, the problem is by no means such an easy one, as these bodies have no satellites. For information in this latter case we have to rely upon such uncertain evidence as, for instance, the slight disturbances caused in the motion of the earth by the attraction of these planets when they pass closest to us, or their observed effect upon the motions of such comets as may happen to pass near to them.

Mass and weight, though often spoken of as one and the same thing, are by no means so. Mass, as we have seen, merely means the amount of matter which a body contains. The weight of a body, on the other hand, depends entirely upon the gravitational pull which it receives. The force of gravity at the surface of the earth is, for instance, about six times as great as that at the surface of the moon. All bodies, therefore, weigh about six times as much on the earth as they would upon the moon; or, rather, a body transferred to the moon's surface would weigh only about one-sixth of what it did on the terrestrial surface. It will therefore be seen that if a body of given mass were to be placed upon planet after planet in turn, its weight would regularly alter according to the force of gravity at each planet's surface.

Gravitation is indeed one of the greatest mysteries of nature. What it is, the means by which it acts, or why such a force should exist at all, are questions to which so far we have not had even the merest hint of an answer. Its action across space appears to be instantaneous.

The intensity of gravitation is said in mathematical parlance "to vary inversely with the square of the distance." This means that at twice the distance the pull will become only one-quarter as strong, and not one-half as otherwise might be expected. At four times the distance, therefore, it will be one-sixteenth as strong. At the earth's surface a body is pulled by the earth's gravitation, or "falls," as we ordinarily term it, through 16 feet in one second of time; whereas at the distance of the moon the attraction of the earth is so very much weakened that a body would take as long as one minute to fall through the same space.

Newton's investigations showed that if a body were to be placed at rest in space entirely away from the attraction of any other body it would remain always in a motionless condition, because there would plainly be no reason why it should move in any one direction rather than in another. And, similarly, if a body were to be projected in a certain direction and at a certain speed, it would move always in the same direction and at the same speed so long as it did not come within the gravitational attraction of any other body.

The possibility of an interaction between the celestial orbs had occurred to astronomers before the time of Newton; for instance, in the ninth century to the Arabian Musa-ben-Shakir, to Camillus Agrippa in 1553, and to Kepler, who suspected its existence from observation of the tides. Horrox also, writing in 1635, spoke of the moon as moved by an emanation from the earth. But no one prior to Newton attempted to examine the question from a mathematical standpoint.

Notwithstanding the acknowledged truth and far-reaching scope of the law of gravitation--for we find its effects exemplified in every portion of the universe--there are yet some minor movements which it does not account for. For instance, there are small irregularities in the movement of Mercury which cannot be explained by the influence of possible intra-Mercurial planets, and similarly there are slight unaccountable deviations in the motions of our neighbour the Moon.

CHAPTER V

CELESTIAL DISTANCES

Up to this we have merely taken a general view of the solar system--a bird's-eye view, so to speak, from space.

In the course of our inquiry we noted in a rough way the relative distances at which the various planets move around the sun. But we have not yet stated what these distances actually are, and it were therefore well now to turn our attention to this important matter.

Each of us has a fair idea of what a mile is. It is a quarter of an hour's sharp walk, for instance; or yonder village or building, we know, lies such and such a number of miles away.

The measurements which have already been given of the diameters of the various bodies of the solar system appear very great to us, who find that a walk of a few miles at a time taxes our strength; but they are a mere nothing when we consider the distances from the sun at which the various planets revolve in their orbits.

The following table gives these distances in round numbers. As here stated they are what are called "mean" distances; for, as the orbits are oval, the planets vary in their distances from the sun, and we are therefore obliged to strike a kind of average for each case:--

Mercury about 36,000,000 miles. Venus " 67,200,000 " Earth " 92,900,000 "

Mars " 141,500,000 " Jupiter " 483,300,000 " Saturn " 886,000,000 " Uranus " 1,781,900,000 " Neptune " 2,791,600,000 "

From the above it will be seen at a glance that we have entered upon a still greater scale of distance than in dealing with the diameters of the various bodies of the system. In that case the distances were limited to thousands of miles; in this, however, we have to deal with millions. A million being ten hundred thousand, it will be noticed that even the diameter of the huge sun is well under a million miles.

How indeed are we to get a grasp of such distances, when those to which we are ordinarily accustomed--the few miles' walk, the little stretch of sea or land which we gaze upon around us--are so utterly minute in comparison? The fact is, that though men may think that they can picture in their minds such immense distances, they actually can not. In matters like these we unconsciously employ a kind of convention, and we estimate a thing as being two or three or more times the size of another. More than this we are unable to do. For instance, our ordinary experience of a mile enables us to judge, in a way, of a stretch of several miles, such as one can take in with a glance; but in our estimation of a thousand miles, or even of one hundred, we are driven back upon a mental trick, so to speak.

In our attempts to realise such immense distances as those in the solar system we are obliged to have recourse to analogies; to comparisons with other and simpler facts, though this is at the best a mere self-cheating device. The analogy which seems most suited to our purpose here, and one which has often been employed by writers, is borrowed from the rate at which an express train travels.

Let us imagine, for instance, that we possess an express train which is capable of running anywhere, never stops, never requires fuel, and always goes along at sixty miles an hour. Suppose we commence by employing it to gauge the size of our own planet, the earth. Let us send it on a trip around the equator, the span of which is about 24,000 miles. At its sixty-miles-an-hour rate of going, this journey will take nearly 17 days. Next let us send it from the earth to the moon. This distance, 240,000 miles, being ten times as great as the last, will of course take ten times as long to cover, namely, 170 days; that is to say, nearly half a year. Again, let us send it still further afield, to the sun, for example. Here, however, it enters upon a journey which is not to be measured in thousands of miles, as the others were, but in millions. The distance from the earth to the sun, as we have seen in the foregoing table, is about 93 millions of miles. Our express train would take about 178 years to traverse this.

Having arrived at the sun, let us suppose that our train makes a tour right round it. This will take more than five years.

Supposing, finally, that our train were started from the sun, and made to run straight out to the known boundaries of the solar system, that is to say, as far as the orbit of Neptune, it would take over 5000 years to traverse the distance.

That sixty miles an hour is a very great speed any one, I think, will admit who has stood upon the platform of a country station while one of the great mail trains has dashed past. But are not the immensities of space appalling to contemplate, when one realises that a body moving incessantly at such a rate would take so long as 10,000 years to traverse merely the breadth of our solar system? Ten thousand years! Just try to conceive it. Why, it is only a little more than half that time since the Pyramids were built, and they mark for us the Dawn of History. And since then half-a-dozen mighty empires have come and gone!

Having thus concluded our general survey of the appearance and dimensions of the solar system, let us next inquire into its position and size in relation to what we call the Universe.

A mere glance at the night sky, when it is free from clouds, shows us that in every direction there are stars; and this holds good, no matter what portion of the globe we visit. The same is really true of the sky by day, though in that case we cannot actually see the stars, for their light is quite overpowered by the dazzling light of the sun.

We thus reach the conclusion that our earth, that our solar system in fact, lies plunged within the midst of a great tangle of stars. What position, by the way, do we occupy in this mighty maze? Are we at the centre, or anywhere near the centre, or where?

It has been indeed amply proved by astronomical research that the stars are bodies giving off a light of their own, just as our sun does; that they are in fact suns, and that our sun is merely one, perhaps indeed a very unimportant member, of this great universe of stars. Each of these stars, or suns, besides, may be the centre of a system similar to what we call our solar system, comprising planets and satellites, comets and meteors;--or perchance indeed some further variety of attendant bodies of which we have no example in our tiny corner of space. But as to whether one is right in a conjecture of this kind, there is up to the present no proof whatever. No telescope has yet shown a planet in attendance upon one of these distant suns; for such bodies, even if they do exist, are entirely out of the range of our mightiest instruments. On what then can we ground such an assumption? Merely upon analogy; upon the common-sense deduction that as the stars have characteristics similar to our particular star, the sun, it would seem unlikely that ours should be the only such body in the whole of space which is attended by a planetary system.

"The Stars," using that expression in its most general sense, do not lie at one fixed distance from us, set here and there upon a background of sky. There is in fact no background at all. The brilliant orbs are all around us in space, at different distances from us and from each other; and we can gaze between them out into the blackness of the void which, perhaps, continues to extend unceasingly long after the very outposts of the stellar universe has been left behind. Shall we then start our imaginary express train once more, and send it out towards the nearest of the stars? This would, however, be a useless experiment. Our express-train method of gauging space would fail miserably in the attempt to bring home to us the mighty gulf by which we are now faced. Let us therefore halt for a moment and look back upon the orders of distance with which we have been dealing. First of all we dealt with thousands of miles. Next we saw how they shrank into insignificance when we embarked upon millions. We found, indeed, that our sixty-mile-an-hour train, rushing along without ceasing, would consume nearly the whole of historical time in a journey from the sun to Neptune.

In the spaces beyond the solar system we are faced, however, by a new order of distance. From sun to planets is measured in millions of miles, but from sun to sun is measured in billions. But does the mere stating of this fact convey anything? I fear not. For the word "billion" runs as glibly off the tongue as "million," and both are so wholly unrealisable by us that the actual difference between them might easily pass unnoticed.

Let us, however, make a careful comparison. What is a million? It is a thousand thousands. But what is a billion? It is a million millions. Consider for a moment! A million of millions. That means a million, each unit of which is again a million. In fact every separate "1" in this million is itself a million. Here is a way of trying to realise this gigantic number. A million seconds make only eleven and a half days and nights. But a billion seconds will make actually more than thirty thousand years!

Having accepted this, let us try and probe with our express train even a little of the new gulf which now lies before us. At our old rate of going it took almost two years to cover a million miles. To cover a billion miles--that is to say, a million times this distance--would thus take of course nearly two million years. Alpha Centauri, the nearest star to our earth, is some twenty-five billions of miles away. Our express train would thus take about fifty millions of years to reach it!

This shows how useless our illustration, appropriate though it seemed for interplanetary space, becomes when applied to the interstellar spaces. It merely gives us millions in return for billions; and so the mind, driven in upon itself, whirls round and round like a squirrel in its revolving cage. There is, however, a useful illustration still left us, and it is the one which astronomers usually employ in dealing with the distances of the stars. The illustration in question is taken from the velocity of light.

Light travels at the tremendous speed of about 186,000 miles a second. It therefore takes only about a second and a quarter to come to us from the moon. It traverses the 93,000,000 of miles which separate us from the sun in about eight minutes. It travels from the sun out to Neptune in about four hours, which means that it would cross the solar system from end to end in eight. To pass, however, across the distance which separates us from Alpha Centauri it would take so long as about four and a quarter years!

Astronomers, therefore, agree in estimating the distances of the stars from the point of view of the time which light would take to pass from them to our earth. They speak of that distance which light takes a year to traverse as a "light year." According to this notation, Alpha Centauri is spoken of as being about four and a quarter light years distant from us.

Now as the rays of light coming from Alpha Centauri to us are chasing one another incessantly across the gulf of space, and as each ray left that star some four years before it reaches us, our view of the star itself must therefore be always some four years old. Were then this star to be suddenly removed from the universe at any moment, we should continue to see it still in its place in the sky for some four years more, after which it would suddenly disappear. The rays which had already started upon their journey towards our earth must indeed continue travelling, and reaching us in their turn until the last one had arrived; after which no more would come.

We have drawn attention to Alpha Centauri as the nearest of the stars. The majority of the others indeed are ever so much farther. We can only hazard a guess at the time it takes for the rays from many of them to reach our globe. Suppose, for instance, we see a sudden change in the light of any of these remote stars, we are inclined to ask ourselves when that change did actually occur. Was it in the days of Queen Elizabeth, or at the time of the Norman Conquest; or was it when Rome was at the height of her glory, or perhaps ages before that when the Pyramids of Egypt were being built? Even the last of these suppositions cannot be treated lightly. We have indeed no real knowledge of the distance from us of those stars which our giant telescopes have brought into view out of the depths of the celestial spaces.

CHAPTER VI

CELESTIAL MEASUREMENT

Had the telescope never been invented our knowledge of astronomy would be trifling indeed.

Prior to the year 1610, when Galileo first turned the new instrument upon the sky, all that men knew of the starry realms was gathered from observation with their own eyes unaided by any artificial means. In such researches they had been very much at a disadvantage. The sun and moon, in their opinion, were no doubt the largest bodies in the heavens, for the mere reason that they looked so! The mighty solar disturbances, which are now such common-places to us, were then quite undreamed of. The moon displayed a patchy surface, and that was all; her craters and ring-mountains were surprises as yet in store for men. Nothing of course was known about the surfaces of the planets. These objects had indeed no particular characteristics to distinguish them from the great host of the stars, except that they continually changed their positions in the sky while the rest did not. The stars themselves were considered as fixed inalterably upon the vault of heaven. The sun, moon, and planets apparently moved about in the intermediate space, supported in their courses by strange and fanciful devices. The idea of satellites was as yet unknown. Comets were regarded as celestial portents, and meteors as small conflagrations taking place in the upper air.

In the entire absence of any knowledge with regard to the actual sizes and distances of the various celestial bodies, men naturally considered them as small; and, concluding that they were comparatively near, assigned to them in consequence a permanent connection with terrestrial affairs. Thus arose the quaint and erroneous beliefs of astrology, according to which the events which took place upon our earth were considered to depend upon the various positions in which the planets, for instance, found themselves from time to time.

It must, however, be acknowledged that the study of astrology, fallacious though its conclusions were, indirectly performed a great service to astronomy by reason of the accurate observations and diligent study of the stars which it entailed.

We will now inquire into the means by which the distances and sizes of the celestial orbs have been ascertained, and see how it was that the ancients were so entirely in the dark in this matter.

There are two distinct methods of finding out the distance at which any object happens to be situated from us.

One method is by actual measurement.

The other is by moving oneself a little to the right or left, and observing whether the distant object appears in any degree altered in position by our own change of place.

One of the best illustrations of this relative change of position which objects undergo as a result of our own change of place, is to observe the landscape from the window of a moving railway carriage. As we are borne rapidly along we notice that the telegraph posts which are set close to the line appear to fly past us in the contrary direction; the trees, houses, and other things beyond go by too, but not so fast; objects a good way off displace slowly; while some spire, or tall landmark, in the far distance appears to remain unmoved during a comparatively long time.

Actual change of position on our own part is found indeed to be invariably accompanied by an apparent displacement of the objects about us, such apparent displacement as a result of our own change of position being known as "parallax." The dependence between the two is so mathematically exact, that if we know the amount of our own change of place, and if we observe the amount of the consequent displacement of any object, we are enabled to calculate its precise distance from us. Thus it comes to pass that distances can be measured without the necessity of moving over them; and the breadth of a river, for instance, or the distance from us of a ship at sea, can be found merely by such means.

It is by the application of this principle to the wider field of the sky that we are able to ascertain the distance of celestial bodies. We have noted that it requires a goodly change of place on our own part to shift the position in which some object in the far distance is seen by us. To two persons separated by, say, a few hundred yards, a ship upon the horizon will appear pretty much in the same direction. They would require, in fact, to be much farther apart in order to displace it sufficiently for the purpose of estimating their distance from it. It is the same with regard to the moon. Two observers, standing upon our earth, will require to be some thousands of miles apart in order to see the position of our satellite sufficiently altered with regard to the starry background, to give the necessary data upon which to ground their calculations.

The change of position thus offered by one side of the earth's surface at a time is, however, not sufficient to displace any but the nearest celestial bodies. When we have occasion to go farther afield we have to seek a greater change of place. This we can get as a consequence of the earth's movement around the sun. Observations, taken several days apart, will show the effect of the earth's change of place during the interval upon the positions of the other bodies of our system. But when we desire to sound the depths of space beyond, and to reach out to measure the distance of the nearest star, we find ourselves at once thrown upon the greatest change of place which we can possibly hope for; and this we get during the long journey of many millions of miles which our earth performs around the sun during the course of each year. But even this last change of place, great as it seems in comparison with terrestrial measurements, is insufficient to show anything more than the tiniest displacements in a paltry forty-three out of the entire host of the stars.

We can thus realise at what a disadvantage the ancients were. The measuring instruments at their command were utterly inadequate to detect such small displacements. It was reserved for the telescope to reveal them; and even then it required the great telescopes of recent times to show the slight changes in the position of the nearer stars, which were caused by the earth's being at one time at one end of its orbit, and some six months later at the other end--stations separated from each other by a gulf of about one hundred and eighty-six millions of miles.

The actual distances of certain celestial bodies being thus ascertainable, it becomes a matter of no great difficulty to determine the actual sizes of the measurable ones. It is a matter of everyday experience that the size which any object appears to have, depends exactly upon the distance it is from us. The farther off it is the smaller it looks; the nearer it is the bigger. If, then, an object which lies at a known distance from us looks such and such a size, we can of course ascertain its real dimensions. Take the moon, for instance. As we have already shown, we are able to ascertain its distance. We observe also that it looks a certain size. It is therefore only a matter of calculation to find what its actual dimensions should be, in order that it may look that size at that distance away. Similarly we can ascertain the real dimensions of the sun. The planets, appearing to us as points of light, seem at first to offer a difficulty; but, by means of the telescope, we can bring them, as it were, so much nearer to us, that their broad expanses may be seen. We fail, however, signally with regard to the stars; for they are so very distant, and therefore such tiny points of light, that our mightiest telescopes cannot magnify them sufficiently to show any breadth of surface.

Instead of saying that an object looks a certain breadth across, such as a yard or a foot, a statement which would really mean nothing, astronomers speak of it as measuring a certain angle. Such angles are estimated in what are called "degrees of arc"; each degree being divided into sixty minutes, and each minute again into sixty seconds. Popularly considered the moon and sun look about the same size, or, as an astronomer would put it, they measure about the same angle. This is an angle, roughly, of thirty-two minutes of arc; that is to say, slightly more than half a degree. The broad expanse of surface which a celestial body shows to us, whether to the naked eye, as in the case of the sun and moon, or in the telescope, as in the case of other members of our system, is technically known as its "disc."

CHAPTER VII

ECLIPSES AND KINDRED PHENOMENA

Since some members of the solar system are nearer to us than others, and all are again much nearer than any of the stars, it must often happen that one celestial body will pass between us and another, and thus intercept its light for a while. The moon, being the nearest object in the universe, will, of course, during its motion across the sky, temporarily blot out every one of the others which happen to lie in its path. When it passes in this manner across the face of the sun, it is said to eclipse it. When it thus hides a planet or star, it is said to occult it. The reason why a separate term is used for what is merely a case of obscuring light in exactly the same way, will be plain when one considers that the disc of the sun is almost of the same apparent size as that of the moon, and so the complete hiding of the sun can last but a few minutes at the most; whereas a planet or a star looks so very small in comparison, that it is always entirely swallowed up for some time when it passes behind the body of our satellite.

The sun, of course, occults planets and stars in exactly the same manner as the moon does, but we cannot see these occultations on account of the blaze of sunlight.

By reason of the small size which the planets look when viewed with the naked eye, we are not able to note them in the act of passing over stars and so blotting them out; but such occurrences may be seen in the telescope, for the planetary bodies then display broad discs.

There is yet another occurrence of the same class which is known as a transit. This takes place when an apparently small body passes across the face of an apparently large one, the phenomenon being in fact the exact reverse of an occultation. As there is no appreciable body nearer to us than the moon, we can never see anything in transit across her disc. But since the planets Venus and Mercury are both nearer to us than the sun, they will occasionally be seen to pass across his face, and thus we get the well-known phenomena called Transits of Venus and Transits of Mercury.

As the satellites of Jupiter are continually revolving around him, they will often pass behind or across his disc. Such occultations and transits of satellites can be well observed in the telescope.

There is, however, a way in which the light of a celestial body may be obscured without the necessity of its being hidden from us by one nearer. It will no doubt be granted that any opaque object casts a shadow when a strong light falls directly upon it. Thus the earth, under the powerful light which is directed upon it from the sun, casts an extensive shadow, though we are not aware of the existence of this shadow until it falls upon something. The shadow which the earth casts is indeed not noticeable to us until some celestial body passes into it. As the sun is very large, and the earth in comparison very small, the shadow thrown by the earth is comparatively short, and reaches out in space for only about a million miles. There is no visible object except the moon, which circulates within that distance from our globe, and therefore she is the only body which can pass into this shadow. Whenever such a thing happens, her surface at once becomes dark, for the reason that she never emits any light of her own, but merely reflects that of the sun. As the moon is continually revolving around the earth, one would be inclined to imagine that once in every month, namely at what is called full moon, when she is on the other side of the earth with respect to the sun, she ought to pass through the shadow in question. But this does not occur every time, because the moon's orbit is not quite upon the same plane with the earth's. It thus happens that time after time the moon passes clear of the earth's shadow, sometimes above it, and sometimes below it. It is indeed only at intervals of about six months that the moon can be thus obscured.

This darkening of her light is known as an eclipse of the moon. It seems a great pity that custom should oblige us to employ the one term "eclipse" for this and also for the quite different occurrence, an eclipse of the sun; in which the sun's face is hidden as a consequence of the moon's body coming directly between it and our eyes.

The popular mind seems always to have found it more difficult to grasp the causes of an eclipse of the moon than an eclipse of the sun. As Mr. J.E. Gore[4] puts it: "The darkening of the sun's light by the interposition of the moon's body seems more obvious than the passing of the moon through the earth's shadow."

Eclipses of the moon furnish striking spectacles, but really add little to our knowledge. They exhibit, however, one of the most remarkable evidences of the globular shape of our earth; for the outline of its shadow when seen creeping over the moon's surface is always circular.

Eclipses of the Moon, or Lunar Eclipses, as they are also called, are of two kinds--Total, and Partial. In a total lunar eclipse the moon passes entirely into the earth's shadow, and the whole of her surface is consequently darkened. This darkening lasts for about two hours. In a partial lunar eclipse, a portion only of the moon passes through the shadow, and so only part of her surface is darkened (see Fig. 3). A very striking phenomenon during a total eclipse of the moon, is that the darkening of the lunar surface is usually by no means so intense as one would expect, when one considers that the sunlight at that time should be wholly cut off from it. The occasions indeed upon which the moon has completely disappeared from view during the progress of a total lunar eclipse are very rare. On the majority of these occasions she has appeared of a coppery-red colour, while sometimes she has assumed an ashen hue. The explanations of these variations of colour is to be found in the then state of the atmosphere which surrounds our earth. When those portions of our earth's atmosphere through which the sun's rays have to filter on their way towards the moon are free from watery vapour, the lunar surface will be tinged with a reddish light, such as we ordinarily experience at sunset when our air is dry. The ashen colour is the result of our atmosphere being laden with watery vapour, and is similar to what we see at sunset when rain is about. Lastly, when the air around the earth is thickly charged with cloud, no light at all can pass; and on such occasions the moon disappears altogether for the time being from the night sky.

Eclipses of the Sun, otherwise known as Solar Eclipses, are divided into Total, Partial, and Annular. A total eclipse of the sun takes place when the moon comes between the sun and the earth, in such a manner that it cuts off the sunlight entirely for the time being from a portion of the earth's surface. A person situated in the region in question will, therefore, at that moment find the sun temporarily blotted out from his view by the body of the moon. Since the moon is a very much smaller body than the sun, and also very much the nearer to us of the two, it will readily be understood that the portion of the earth from which the sun is seen thus totally eclipsed will be of small extent. In places not very distant from this region, the moon will appear so much shifted in the sky that the sun will be seen only partially eclipsed. The moon being in constant movement round the earth, the portion of the earth's surface from which an eclipse is seen as total will be always a comparatively narrow band lying roughly from west to east. This band, known as the track of totality, can, at the utmost, never be more than about 165 miles in width, and as a rule is very much less. For about 2000 miles on either side of it the sun is seen partially eclipsed. Outside these limits no eclipse of any kind is visible, as from such regions the moon is not seen to come in the way of the sun (see Fig. 4 (i.), p. 67).

It may occur to the reader that eclipses can also take place in the course of which the positions, where the eclipse would ordinarily be seen as total, will lie outside the surface of the earth. Such an eclipse is thus not dignified with the name of total eclipse, but is called a partial eclipse, because from the earth's surface the sun is only seen partly eclipsed at the utmost (see Fig. 4 (ii.), p. 67).

An Annular eclipse is an eclipse which just fails to become total for yet another reason. We have pointed out that the orbits of the various members of the solar system are not circular, but oval. Such oval figures, it will be remembered, are technically known as ellipses. In an elliptic orbit the controlling body is situated not in the middle of the figure, but rather towards one of the ends; the actual point which it occupies being known as the focus. The sun being at the focus of the earth's orbit, it follows that the earth is, at times, a little nearer to him than at others. The sun will therefore appear to us to vary a little in size, looking sometimes slightly larger than at other times.

It is so, too, with the moon, at the focus of whose orbit the earth is situated. She therefore also appears to us at times to vary slightly in size. The result is that when the sun is eclipsed by the moon, and the moon at the time appears the larger of the two, she is able to blot out the sun completely, and so we can get a total eclipse. But when, on the other hand, the sun appears the larger, the eclipse will not be quite total, for a portion of the sun's disc will be seen protruding all around the moon like a ring of light. This is what is known as an annular eclipse, from the Latin word annulus, which means a ring. The term is consecrated by long usage, but it seems an unfortunate one on account of its similarity to the word "annual." The Germans speak of this kind of eclipse as "ring-formed," which is certainly much more to the point.

There can never be a year without an eclipse of the sun. Indeed there must be always two such eclipses at least during that period, though there need be no eclipse of the moon at all. On the other hand, the greatest number of eclipses which can ever take place during a year are seven; that is to say, either five solar eclipses and two lunar, or four solar and three lunar. This general statement refers merely to eclipses in their broadest significance, and informs us in no way whether they will be total or partial.

Of all the phenomena which arise from the hiding of any celestial body by one nearer coming in the way, a total eclipse of the sun is far the most important. It is, indeed, interesting to consider how much poorer modern astronomy would be but for the extraordinary coincidence which makes a total solar eclipse just possible. The sun is about 400 times farther off from us than the moon, and enormously greater than her in bulk. Yet the two are relatively so distanced from us as to look about the same size. The result of this is that the moon, as has been seen, can often blot out the sun entirely from our view for a short time. When this takes place the great blaze of sunlight which ordinarily dazzles our eyes is completely cut off, and we are thus enabled, unimpeded, to note what is going on in the immediate vicinity of the sun itself.

In a total solar eclipse, the time which elapses from the moment when the moon's disc first begins to impinge upon that of the sun at his western edge until the eclipse becomes total, lasts about an hour. During all this time the black lunar disc may be watched making its way steadily across the solar face. Notwithstanding the gradual obscuration of the sun, one does not notice much diminution of light until about three-quarters of his disc are covered. Then a wan, unearthly appearance begins to pervade all things, the temperature falls noticeably, and nature seems to halt in expectation of the coming of something unusual. The decreasing portion of sun becomes more and more narrow, until at length it is reduced to a crescent-shaped strip of exceeding fineness. Strange, ill-defined, flickering shadows (known as "Shadow Bands") may at this moment be seen chasing each other across any white expanse such as a wall, a building, or a sheet stretched upon the ground. The western side of the sky has now assumed an appearance dark and lowering, as if a rainstorm of great violence were approaching. This is caused by the mighty mass of the lunar shadow sweeping rapidly along. It flies onward at the terrific velocity of about half a mile a second.

If the gradually diminishing crescent of sun be now watched through a telescope, the observer will notice that it does not eventually vanish all at once, as he might have expected. Rather, it breaks up first of all along its length into a series of brilliant dots, known as "Baily's Beads." The reason of this phenomenon is perhaps not entirely agreed upon, but the majority of astronomers incline to the opinion that the so-called "beads" are merely the last remnants of sunlight peeping between those lunar mountain peaks which happen at the moment to fringe the advancing edge of the moon. The beads are no sooner formed than they rapidly disappear one after the other, after which no portion of the solar surface is left to view, and the eclipse is now total (see Fig. 5).

But with the disappearance of the sun there springs into view a new and strange appearance, ordinarily unseen because of the blaze of sunlight. It is a kind of aureole, or halo, pearly white in colour, which is seen to surround the black disc of the moon. This white radiance is none other than the celebrated phenomenon widely known as the Solar Corona. It was once upon a time thought to belong to the moon, and to be perhaps a lunar atmosphere illuminated by the sunlight shining through it from behind. But the suddenness with which the moon always blots out stars when occulting them, has amply proved that she possesses no atmosphere worth speaking about. It is now, however, satisfactorily determined that the corona belongs to the sun, for during the short time that it remains in view the black body of the moon can be seen creeping across it.

All the time that the total phase (as it is called) lasts, the corona glows with its pale unearthly light, shedding upon the earth's surface an illumination somewhat akin to full moonlight. Usually the planet Venus and a few stars shine out the while in the darkened heaven. Meantime around the observer animal and plant life behave as at nightfall. Birds go to roost, bats fly out, worms come to the surface of the ground, flowers close up. In the Norwegian eclipse of 1896 fish were seen rising to the surface of the water. When the total phase at length is over, and the moon in her progress across the sky has allowed the brilliant disc of the sun to spring into view once more at the other side, the corona disappears.

There is another famous accompaniment of the sun which partly reveals itself during total solar eclipses. This is a layer of red flame which closely envelops the body of the sun and lies between it and the corona. This layer is known by the name of the Chromosphere. Just as at ordinary times we cannot see the corona on account of the blaze of sunlight, so are we likewise unable to see the chromosphere because of the dazzling white light which shines through from the body of the sun underneath and completely overpowers it. When, however, during a solar eclipse, the lunar disc has entirely hidden the brilliant face of the sun, we are still able for a few moments to see an edgewise portion of the chromosphere in the form of a narrow red strip, fringing the advancing border of the moon. Later on, just before the moon begins to uncover the face of the sun from the other side, we may again get a view of a strip of chromosphere.

The outer surface of the chromosphere is not by any means even. It is rough and billowy, like the surface of a storm-tossed sea. Portions of it, indeed, rise at times to such heights that they may be seen standing out like blood-red points around the black disc of the moon, and remain thus during a good part of the total phase. These projections are known as the Solar Prominences. In the same way as the corona, the chromosphere and prominences were for a time supposed to belong to the moon. This, however, was soon found not to be the case, for the lunar disc was noticed to creep slowly across them also.

The total phase, or "totality," as it is also called, lasts for different lengths of time in different eclipses. It is usually of about two or three minutes' duration, and at the utmost it can never last longer than about eight minutes.

When totality is over and the corona has faded away, the moon's disc creeps little by little from the face of the sun, light and heat returns once more to the earth, and nature recovers gradually from the gloom in which she has been plunged. About an hour after totality, the last remnant of moon draws away from the solar disc, and the eclipse is entirely at an end.

The corona, the chromosphere, and the prominences are the most important of these accompaniments of the sun which a total eclipse reveals to us. Our further consideration of them must, however, be reserved for a subsequent chapter, in which the sun will be treated of at length.

Every one who has had the good fortune to see a total eclipse of the sun will, the writer feels sure, agree with the verdict of Sir Norman Lockyer that it is at once one of the "grandest and most awe-inspiring sights" which man can witness. Needless to say, such an occurrence used to cause great consternation in less civilised ages; and that it has not in modern times quite parted with its terrors for some persons, is shown by the fact that in Iowa, in the United States, a woman died from fright during the eclipse of 1869.

To the serious observer of a total solar eclipse every instant is extremely precious. Many distinct observations have to be crowded into a time all too limited, and this in an eclipse-party necessitates constant rehearsals in order that not a moment may be wasted when the longed-for totality arrives. Such preparation is very necessary; for the rarity and uncommon nature of a total eclipse of the sun, coupled with its exceeding short duration, tends to flurry the mind, and to render it slow to seize upon salient points of detail. And, even after every precaution has been taken, weather possibilities remain to be reckoned with, so that success is rather a lottery.

Above all things, therefore, a total solar eclipse is an occurrence for the proper utilisation of which personal experience is of absolute necessity. It was manifestly out of the question that such experience could be gained by any individual in early times, as the imperfection of astronomical theory and geographical knowledge rendered the predicting of the exact position of the track of totality well-nigh impossible. Thus chance alone would have enabled one in those days to witness a total phase, and the probabilities, of course, were much against a second such experience in the span of a life-time. And even in more modern times, when the celestial motions had come to be better understood, the difficulties of foreign travel still were in the way; for it is, indeed, a notable fact that during many years following the invention of the telescope the tracks were placed for the most part in far-off regions of the earth, and Europe was visited by singularly few total solar eclipses. Thus it came to pass that the building up of a body of organised knowledge upon this subject was greatly delayed.

Nothing perhaps better shows the soundness of modern astronomical theory than the almost exact agreement of the time predicted for an eclipse with its actual occurrence. Similarly, by calculating backwards, astronomers have discovered the times and seasons at which many ancient eclipses took place, and valuable opportunities have thus arisen for checking certain disputed dates in history.

It should not be omitted here that the ancients were actually able, in a rough way, to predict eclipses. The Chaldean astronomers had indeed noticed very early a curious circumstance, i.e. that eclipses tend to repeat themselves after a lapse of slightly more than eighteen years.

In this connection it must, however, be pointed out, in the first instance, that the eclipses which occur in any particular year are in no way associated with those which occurred in the previous year. In other words, the mere fact that an eclipse takes place upon a certain day this year will not bring about a repetition of it at the same time next year. However, the nicely balanced behaviour of the solar system, an equilibrium resulting from orbital ebb and flow, naturally tends to make the members which compose that family repeat their ancient combinations again and again; so that after definite lapses of time the same order of things will almost exactly recur. Thus, as a consequence of their beautifully poised motions, the sun, the moon, and the earth tend, after a period of 18 years and 10-1/3 days,[5] to occupy very nearly the same positions with regard to each other. The result of this is that, during each recurring period, the eclipses comprised within it will be repeated in their order.

To give examples:--

The total solar eclipse of August 30, 1905, was a repetition of that of August 19, 1887.

The partial solar eclipse of February 23, 1906, corresponded to that which took place on February 11, 1888.

The annular eclipse of July 10, 1907, was a recurrence of that of June 28, 1889.

In this way we can go on until the eighteen year cycle has run out, and we come upon a total solar eclipse predicted for September 10, 1923, which will repeat the above-mentioned ones of 1905 and 1887; and so on too with the others.

From mere observation alone, extending no doubt over many ages, those time-honoured watchers of the sky, the early Chaldeans, had arrived at this remarkable generalisation; and they used it for the rough prediction of eclipses. To the period of recurrence they give the name of "Saros."

And here we find ourselves led into one of the most interesting and fascinating by-paths in astronomy, to which writers, as a rule, pay all too little heed.

In order not to complicate matters unduly, the recurrence of solar eclipses alone will first be dealt with. This limitation will, however, not affect the arguments in the slightest, and it will be all the more easy in consequence to show their application to the case of eclipses of the moon.

The reader will perhaps have noticed that, with regard to the repetition of an eclipse, it has been stated that the conditions which bring it on at each recurrence are reproduced almost exactly. Here, then, lies the crux of the situation. For it is quite evident that were the conditions exactly reproduced, the recurrences of each eclipse would go on for an indefinite period. For instance, if the lapse of a saros period found the sun, moon, and earth again in the precise relative situations which they had previously occupied, the recurrences of a solar eclipse would tend to duplicate its forerunner with regard to the position of the shadow upon the terrestrial surface. But the conditions not being exactly reproduced, the shadow-track does not pass across the earth in quite the same regions. It is shifted a little, so to speak; and each time the eclipse comes round it is found to be shifted a little farther.

Every solar eclipse has therefore a definite "life" of its own upon the earth, lasting about 1150 years, or 64 saros returns, and working its way little by little across our globe from north to south, or from south to north, as the case may be. Let us take an imaginary example. A partial eclipse occurs, say, somewhere near the North Pole, the edge of the "partial" shadow just grazing the earth, and the "track of totality" being as yet cast into space. Here we have the beginning of a series. At each saros recurrence the partial shadow encroaches upon a greater extent of earth-surface. At length, in its turn, the track of totality begins to impinge upon the earth. This track streaks across our globe at each return of the eclipse, repeating itself every time in a slightly more southerly latitude. South and south it moves, passing in turn the Tropic of Cancer, the Equator, the Tropic of Capricorn, until it reaches the South Pole; after which it touches the earth no longer, but is cast into space. The rear portion of the partial shadow, in its turn, grows less and less in extent; and it too in time finally passes off. Our imaginary eclipse series is now no more--its "life" has ended.

We have taken, as an example, an eclipse series moving from north to south. We might have taken one moving from south to north, for they progress in either direction.

From the description just given the reader might suppose that, if the tracks of totality of an eclipse series were plotted upon a chart of the world, they would lie one beneath another like a set of steps. This is, however, not the case, and the reason is easily found. It depends upon the fact that the saros does not comprise an exact number of days, but includes, as we have seen, one-third of a day in addition.

It will be granted, of course, that if the number of days was exact, the same parts of the earth would always be brought round by the axial rotation to front the sun at the moment of the recurrence of the eclipse. But as there is still one-third of a day to complete the saros period, the earth has yet to make one-third of a rotation upon its axis before the eclipse takes place. Thus at every recurrence the track of totality finds itself placed one-third of the earth's circumference to the westward. Three of the recurrences will, of course, complete the circuit of the globe; and so the fourth recurrence will duplicate the one which preceded it, three saros returns, or 54 years and 1 month before. This duplication, as we have already seen, will, however, be situated in a latitude to the south or north of its predecessor, according as the eclipse series is progressing in a southerly or northerly direction.

Lastly, every eclipse series, after working its way across the earth, will return again to go through the same process after some 12,000 years; so that, at the end of that great lapse of time, the entire "life" of every eclipse should repeat itself, provided that the conditions of the solar system have not altered appreciably during the interval.

We are now in a position to consider this gradual southerly or northerly progress of eclipse recurrences in its application to the case of eclipses of the moon. It should be evident that, just as in solar eclipses the lunar shadow is lowered or raised (as the case may be) each time it strikes the terrestrial surface, so in lunar eclipses will the body of the moon shift its place at each recurrence relatively to the position of the earth's shadow. Every lunar eclipse, therefore, will commence on our satellite's disc as a partial eclipse at the northern or southern extremity, as the case may be. Let us take, as an example, an imaginary series of eclipses of the moon progressing from north to south. At each recurrence the partial phase will grow greater, its boundary encroaching more and more to the southward, until eventually the whole disc is enveloped by the shadow, and the eclipse becomes total. It will then repeat itself as total during a number of recurrences, until the entire breadth of the shadow has been passed through, and the northern edge of the moon at length springs out into sunlight. This illuminated portion will grow more and more extensive at each succeeding return, the edge of the shadow appearing to recede from it until it finally passes off at the south. Similarly, when a lunar eclipse commences as partial at the south of the moon, the edge of the shadow at each subsequent recurrence finds itself more and more to the northward. In due course the total phase will supervene, and will persist during a number of recurrences until the southerly trend of the moon results in the uncovering of the lunar surface at the south. Thus, as the boundary of the shadow is left more and more to the northward, the illuminated portion on the southern side of the moon becomes at each recurrence greater and the darkened portion on the northern side less, until the shadow eventually passes off at the north.

The "life" of an eclipse of the moon happens, for certain reasons, to be much shorter than that of an eclipse of the sun. It lasts during only about 860 years, or 48 saros returns.

Fig. 6, p. 81, is a map of the world on Mercator's Projection, showing a portion of the march of the total solar eclipse of August 30, 1905, across the surface of the earth. The projection in question has been employed because it is the one with which people are most familiar. This eclipse began by striking the neighbourhood of the North Pole in the guise of a partial eclipse during the latter part of the reign of Queen Elizabeth, and became total on the earth for the first time on the 24th of June 1797. Its next appearance was on the 6th of July 1815. It has not been possible to show the tracks of totality of these two early visitations on account of the distortion of the polar regions consequent on the fiction of Mercator's Projection. It is therefore made to commence with the track of its third appearance, viz. on July 17, 1833. In consequence of those variations in the apparent sizes of the sun and moon, which result, as we have seen, from the variations in their distances from the earth, this eclipse will change from a total into an annular eclipse towards the end of the twenty-first century. By that time the track will have passed to the southern side of the equator. The track will eventually leave the earth near the South Pole about the beginning of the twenty-sixth century, and the rear portion of the partial shadow will in its turn be clear of the terrestrial surface by about 2700 A.D., when the series comes to an end.

[4] Astronomical Essays (p. 40), London, 1907.

[5] In some cases the periods between the dates of the corresponding eclipses appear to include a greater number of days than ten; but this is easily explained when allowance is made for intervening leap years (in each of which an extra day has of course been added), and also for variations in local time.

CHAPTER VIII

FAMOUS ECLIPSES OF THE SUN

What is thought to be the earliest reference to an eclipse comes down to us from the ancient Chinese records, and is over four thousand years old. The eclipse in question was a solar one, and occurred, so far as can be ascertained, during the twenty-second century B.C. The story runs that the two state astronomers, Ho and Hi by name, being exceedingly intoxicated, were unable to perform their required duties, which consisted in superintending the customary rites of beating drums, shooting arrows, and the like, in order to frighten away the mighty dragon which it was believed was about to swallow up the Lord of Day. This eclipse seems to have been only partial; nevertheless a great turmoil ensued, and the two astronomers were put to death, no doubt with the usual celestial cruelty.

The next eclipse mentioned in the Chinese annals is also a solar eclipse, and appears to have taken place more than a thousand years later, namely in 776 B.C. Records of similar eclipses follow from the same source; but as they are mere notes of the events, and do not enter into any detail, they are of little interest. Curiously enough the Chinese have taken practically no notice of eclipses of the moon, but have left us a comparatively careful record of comets, which has been of value to modern astronomy.

The earliest mention of a total eclipse of the sun (for it should be noted that the ancient Chinese eclipse above-mentioned was merely partial) was deciphered in 1905, on a very ancient Babylonian tablet, by Mr. L.W. King of the British Museum. This eclipse took place in the year 1063 B.C.

Assyrian tablets record three solar eclipses which occurred between three and four hundred years later than this. The first of these was in 763 B.C.; the total phase being visible near Nineveh.

The next record of an eclipse of the sun comes to us from a Grecian source. This eclipse took place in 585 B.C., and has been the subject of much investigation. Herodotus, to whom we are indebted for the account, tells us that it occurred during a battle in a war which had been waging for some years between the Lydians and Medes. The sudden coming on of darkness led to a termination of the contest, and peace was afterwards made between the combatants. The historian goes on to state that the eclipse had been foretold by Thales, who is looked upon as the Founder of Grecian astronomy. This eclipse is in consequence known as the "Eclipse of Thales." It would seem as if that philosopher were acquainted with the Chaldean saros.

The next solar eclipse worthy of note was an annular one, and occurred in 431 B.C., the first year of the Peloponnesian War. Plutarch relates that the pilot of the ship, which was about to convey Pericles to the Peloponnesus, was very much frightened by it; but Pericles calmed him by holding up a cloak before his eyes, and saying that the only difference between this and the eclipse was that something larger than the cloak prevented his seeing the sun for the time being.

An eclipse of great historical interest is that known as the "Eclipse of Agathocles," which occurred on the morning of the 15th of August, 310 B.C. Agathocles, Tyrant of Syracuse, had been blockaded in the harbour of that town by the Carthaginian fleet, but effected the escape of his squadron under cover of night, and sailed for Africa in order to invade the enemy's territory. During the following day he and his vessels experienced a total eclipse, in which "day wholly put on the appearance of night, and the stars were seen in all parts of the sky."

A few solar eclipses are supposed to be referred to in early Roman history, but their identity is very doubtful in comparison with those which the Greeks have recorded. Additional doubt is cast upon them by the fact that they are usually associated with famous events. The birth and death of Romulus, and the Passage of the Rubicon by Julius Caesar, are stated indeed to have been accompanied by these marks of the approval or disapproval of the gods!

Reference to our subject in the Bible is scanty. Amos viii. 9 is thought to refer to the Nineveh eclipse of 763 B.C., to which allusion has already been made; while the famous episode of Hezekiah and the shadow on the dial of Ahaz has been connected with an eclipse which was partial at Jerusalem in 689 B.C.

The first solar eclipse, recorded during the Christian Era, is known as the "Eclipse of Phlegon," from the fact that we are indebted for the account to a pagan writer of that name. This eclipse took place in A.D. 29, and the total phase was visible a little to the north of Palestine. It has sometimes been confounded with the "darkness of the Crucifixion," which event took place near the date in question; but it is sufficient here to say that the Crucifixion is well known to have occurred during the Passover of the Jews, which is always celebrated at the full moon, whereas an eclipse of the sun can only take place at new moon.

Dion Cassius, commenting on the Emperor Claudius about the year A.D. 45, writes as follows:--

"As there was going to be an eclipse on his birthday, through fear of a disturbance, as there had been other prodigies, he put forth a public notice, not only that the obscuration would take place, and about the time and magnitude of it, but also about the causes that produce such an event."

This is a remarkable piece of information; for the Romans, an essentially military nation, appear hitherto to have troubled themselves very little about astronomical matters, and were content, as we have seen, to look upon phenomena, like eclipses, as mere celestial prodigies.

What is thought to be the first definite mention of the solar corona occurs in a passage of Plutarch. The eclipse to which he refers is probably one which took place in A.D. 71. He says that the obscuration caused by the moon "has no time to last and no extensiveness, but some light shows itself round the sun's circumference, which does not allow the darkness to become deep and complete." No further reference to this phenomenon occurs until near the end of the sixteenth century. It should, however, be here mentioned that Mr. E.W. Maunder has pointed out the probability[6] that we have a very ancient symbolic representation of the corona in the "winged circle," "winged disc," or "ring with wings," as it is variously called, which appears so often upon Assyrian and Egyptian monuments, as the symbol of the Deity (Fig. 7).

[Illustration: FIG. 7.--The "Ring with Wings." The upper is the Assyrian form of the symbol, the lower the Egyptian. (From Knowledge.) Compare the form of the corona on Plate VII. (B), p. 142.]

The first solar eclipse recorded to have been seen in England is that of A.D. 538, mention of which is found in the Anglo-Saxon Chronicle. The track of totality did not, however, come near our islands, for only two-thirds of the sun's disc were eclipsed at London.

In 840 a great eclipse took place in Europe, which was total for more than five minutes across what is now Bavaria. Terror at this eclipse is said to have hastened the death of Louis le Debonnaire, Emperor of the West, who lay ill at Worms.

In 878--temp. King Alfred--an eclipse of the sun took place which was total at London. From this until 1715 no other eclipse was total at London itself; though this does not apply to other portions of England.

An eclipse, generally known as the "Eclipse of Stiklastad," is said to have taken place in 1030, during the sea-fight in which Olaf of Norway is supposed to have been slain. Longfellow, in his Saga of King Olaf, has it that

"The Sun hung red As a drop of blood,"

but, as in the case of most poets, the dramatic value of an eclipse seems to have escaped his notice.

In the year 1140 there occurred a total eclipse of the sun, the last to be visible in England for more than five centuries. Indeed there have been only two such since--namely, those of 1715 and 1724, to which we shall allude in due course. The eclipse of 1140 took place on the 20th March, and is thus referred to in the Anglo-Saxon Chronicle:--

"In the Lent, the sun and the day darkened, about the noon-tide of the day, when men were eating, and they lighted candles to eat by. That was the 13th day before the calends of April. Men were very much struck with wonder."

Several of the older historians speak of a "fearful eclipse" as having taken place on the morning of the Battle of Crecy, 1346. Lingard, for instance, in his History of England, has as follows:--

"Never, perhaps, were preparations for battle made under circumstances so truly awful. On that very day the sun suffered a partial eclipse: birds, in clouds, the precursors of a storm, flew screaming over the two armies, and the rain fell in torrents, accompanied by incessant thunder and lightning. About five in the afternoon the weather cleared up; the sun in full splendour darted his rays in the eyes of the enemy."

Calculations, however, show that no eclipse of the sun took place in Europe during that year. This error is found to have arisen from the mistranslation of an obsolete French word esclistre (lightning), which is employed by Froissart in his description of the battle.

In 1598 an eclipse was total over Scotland and part of North Germany. It was observed at Torgau by Jessenius, an Hungarian physician, who noticed a bright light around the moon during the time of totality. This is said to be the first reference to the corona since that of Plutarch, to which we have already drawn attention.

Mention of Scotland recalls the fact that an unusual number of eclipses happen to have been visible in that country, and the occult bent natural to the Scottish character has traditionalised a few of them in such terms as the "Black Hour" (an eclipse of 1433), "Black Saturday" (the eclipse of 1598 which has been alluded to above), and "Mirk Monday" (1652). The track of the last-named also passed over Carrickfergus in Ireland, where it was observed by a certain Dr. Wybord, in whose account the term "corona" is first employed. This eclipse is the last which has been total in Scotland, and it is calculated that there will not be another eclipse seen as total there until the twenty-second century.

An eclipse of the sun which took place on May 30, 1612, is recorded as having been seen "through a tube." This probably refers to the then recent invention--the telescope.

The eclipses which we have been describing are chiefly interesting from an historical point of view. The old mystery and confusion to the beholders seem to have lingered even into comparatively enlightened times, for we see how late it is before the corona attracts definite attention for the sake of itself alone.

It is not a far cry from notice of the corona to that of other accompaniments of a solar eclipse. Thus the eclipse of 1706, the total phase of which was visible in Switzerland, is of great interest; for it was on this occasion that the famous red prominences seem first to have been noted. A certain Captain Stannyan observed this eclipse from Berne in Switzerland, and described it in a letter to Flamsteed, the then Astronomer Royal. He says the sun's "getting out of his eclipse was preceded by a blood-red streak of light from its left limb, which continued not longer than six or seven seconds of time; then part of the Sun's disc appeared all of a sudden, as bright as Venus was ever seen in the night, nay brighter; and in that very instant gave a Light and Shadow to things as strong as Moonlight uses to do." How little was then expected of the sun is, however, shown by Flamsteed's words, when communicating this information to the Royal Society:--

"The Captain is the first man I ever heard of that took notice of a Red Streak of Light preceding the Emersion of the Sun's body from a total Eclipse. And I take notice of it to you because it infers that the Moon has an atmosphere; and its short continuance of only six or seven seconds of time, tells us that its height is not more than the five or six hundredth part of her diameter."

What a change has since come over the ideas of men! The sun has proved a veritable mine of discovery, while the moon has yielded up nothing new.

The eclipse of 1715, the first total at London since that of 878, was observed by the famous astronomer, Edmund Halley, from the rooms of the Royal Society, then in Crane Court, Fleet Street. On this occasion both the corona and a red projection were noted. Halley further makes allusion to that curious phenomenon, which later on became celebrated under the name of "Baily's beads." It was also on the occasion of this eclipse that the earliest recorded drawings of the corona were made. Cambridge happened to be within the track of totality; and a certain Professor Cotes of that University, who is responsible for one of the drawings in question, forwarded them to Sir Isaac Newton together with a letter describing his observations.

In 1724 there occurred an eclipse, the total phase of which was visible from the south-west of England, but not from London. The weather was unfavourable, and the eclipse consequently appears to have been seen by only one person, a certain Dr. Stukeley, who observed it from Haraden Hill near Salisbury Plain. This is the last eclipse of which the total phase was seen in any part of England. The next will not be until June 29, 1927, and will be visible along a line across North Wales and Lancashire. The discs of the sun and moon will just then be almost of the same apparent size, and so totality will be of extremely short duration; in fact only a few seconds. London itself will not see a totality until the year 2151--a circumstance which need hardly distress any of us personally!

It is only from the early part of the nineteenth century that serious scientific attention to eclipses of the sun can be dated. An annular eclipse, visible in 1836 in the south of Scotland, drew the careful notice of Francis Baily of Jedburgh in Roxburghshire to that curious phenomenon which we have already described, and which has ever since been known by the name of "Baily's beads." Spurred by his observation, the leading astronomers of the day determined to pay particular attention to a total eclipse, which in the year 1842 was to be visible in the south of France and the north of Italy. The public interest aroused on this occasion was also very great, for the region across which the track of totality was to pass was very populous, and inhabited by races of a high degree of culture.

This eclipse occurred on the morning of the 8th July, and from it may be dated that great enthusiasm with which total eclipses of the sun have ever since been received. Airy, our then Astronomer Royal, observed it from Turin; Arago, the celebrated director of the Paris Observatory, from Perpignan in the south of France; Francis Baily from Pavia; and Sir John Herschel from Milan. The corona and three large red prominences were not only well observed by the astronomers, but drew tremendous applause from the watching multitudes.

The success of the observations made during this eclipse prompted astronomers to pay similar attention to that of July 28, 1851, the total phase of which was to be visible in the south of Norway and Sweden, and across the east of Prussia. This eclipse was also a success, and it was now ascertained that the red prominences belonged to the sun and not to the moon; for the lunar disc, as it moved onward, was seen to cover and to uncover them in turn. It was also noted that these prominences were merely uprushes from a layer of glowing gaseous matter, which was seen closely to envelop the sun.

The total eclipse of July 18, 1860, was observed in Spain, and photography was for the first time systematically employed in its observation.[7] In the photographs taken the stationary appearance of both the corona and prominences with respect to the moving moon, definitely confirmed the view already put forward that they were actual appendages of the sun.

The eclipse of August 18, 1868, the total phase of which lasted nearly six minutes, was visible in India, and drew thither a large concourse of astronomers. In this eclipse the spectroscope came to the front, and showed that both the prominences, and the chromospheric layer from which they rise, are composed of glowing vapours--chief among which is the vapour of hydrogen. The direct result of the observations made on this occasion was the spectroscopic method of examining prominences at any time in full daylight, and without a total eclipse. This method, which has given such an immense impetus to the study of the sun, was the outcome of independent and simultaneous investigation on the part of the French astronomer, the late M. Janssen, and the English astronomer, Professor (now Sir Norman) Lockyer, a circumstance strangely reminiscent of the discovery of Neptune. The principles on which the method was founded seem, however, to have occurred to Dr. (now Sir William) Huggins some time previously.

The eclipse of December 22, 1870, was total for a little more than two minutes, and its track passed across the Mediterranean. M. Janssen, of whom mention has just been made, escaped in a balloon from then besieged Paris, taking his instruments with him, and made his way to Oran, in Algeria, in order to observe it; but his expectations were disappointed by cloudy weather. The expedition sent out from England had the misfortune to be shipwrecked off the coast of Sicily. But the occasion was redeemed by a memorable observation made by the American astronomer, the late Professor Young, which revealed the existence of what is now known as the "Reversing Layer." This is a shallow layer of gases which lies immediately beneath the chromosphere. An illustration of the corona, as it was seen during the above eclipse, will be found on Plate VII. (A), p. 142.

In the eclipse of December 12, 1871, total across Southern India, the photographs of the corona obtained by Mr. Davis, assistant to Lord Lindsay (now the Earl of Crawford), displayed a wealth of detail hitherto unapproached.

The eclipse of July 29, 1878, total across the western states of North America, was a remarkable success, and a magnificent view of the corona was obtained by the well-known American astronomer and physicist, the late Professor Langley, from the summit of Pike's Peak, Colorado, over 14,000 feet above the level of the sea. The coronal streamers were seen to extend to a much greater distance at this altitude than at points less elevated, and the corona itself remained visible during more than four minutes after the end of totality. It was, however, not entirely a question of altitude; the coronal streamers were actually very much longer on this occasion than in most of the eclipses which had previously been observed.

The eclipse of May 17, 1882, observed in Upper Egypt, is notable from the fact that, in one of the photographs taken by Dr. Schuster at Sohag, a bright comet appeared near the outer limit of the corona (see Plate I., p. 96). The comet in question had not been seen before the eclipse, and was never seen afterwards. This is the third occasion on which attention has been drawn to a comet merely by a total eclipse. The first is mentioned by Seneca; and the second by Philostorgius, in an account of an eclipse observed at Constantinople in A.D. 418. A fourth case of the kind occurred in 1893, when faint evidences of one of these filmy objects were found on photographs of the corona taken by the American astronomer, Professor Schaeberle, during the total eclipse of April 16 of that year.

The eclipse of May 6, 1883, had a totality of over five minutes, but the central track unfortunately passed across the Pacific Ocean, and the sole point of land available for observing it from was one of the Marquesas Group, Caroline Island, a coral atoll seven and a half miles long by one and a half broad. Nevertheless astronomers did not hesitate to take up their posts upon that little spot, and were rewarded with good weather.

The next eclipse of importance was that of April 16, 1893. It stretched from Chili across South America and the Atlantic Ocean to the West Coast of Africa, and, as the weather was fine, many good results were obtained. Photographs were taken at both ends of the track, and these showed that the appearance of the corona remained unchanged during the interval of time occupied by the passage of the shadow across the earth. It was on the occasion of this eclipse that Professor Schaeberle found upon his photographs those traces of the presence of a comet, to which allusion has already been made.

Extensive preparations were made to observe the eclipse of August 9, 1896. Totality lasted from two to three minutes, and the track stretched from Norway to Japan. Bad weather disappointed the observers, with the exception of those taken to Nova Zembla by Sir George Baden Powell in his yacht Otaria.

The eclipse of January 22, 1898, across India vi?Bombay and Benares, was favoured with good weather, and is notable for a photograph obtained by Mrs. E.W. Maunder, which showed a ray of the corona extending to a most unusual distance.

Of very great influence in the growth of our knowledge with regard to the sun, is the remarkable piece of good fortune by which the countries around the Mediterranean, so easy of access, have been favoured with a comparatively large number of total eclipses during the past sixty years. Tracks of totality have, for instance, traversed the Spanish peninsula on no less than five occasions during that period. Two of these are among the most notable eclipses of recent years, namely, those of May 28, 1900, and of August 30, 1905. In the former the track of totality stretched from the western seaboard of Mexico, through the Southern States of America, and across the Atlantic Ocean, after which it passed over Portugal and Spain into North Africa. The total phase lasted for about a minute and a half, and the eclipse was well observed from a great many points along the line. A representation of the corona, as it appeared on this occasion, will be found on Plate VII. (B), p. 142.

The track of the other eclipse to which we have alluded, i.e. that of August 30, 1905, crossed Spain about 200 miles to the northward of that of 1900. It stretched from Winnipeg in Canada, through Labrador, and over the Atlantic; then traversing Spain, it passed across the Balearic Islands, North Africa, and Egypt, and ended in Arabia (see Fig. 6, p. 81). Much was to be expected from a comparison between the photographs taken in Labrador and Egypt on the question as to whether the corona would show any alteration in shape during the time that the shadow was traversing the intervening space--some 6000 miles. The duration of the total phase in this eclipse was nearly four minutes. Bad weather, however, interfered a good deal with the observations. It was not possible, for instance, to do anything at all in Labrador. In Spain the weather conditions were by no means favourable; though at Burgos, where an immense number of people had assembled, the total phase was, fortunately, well seen. On the whole, the best results were obtained at Guelma in Algeria. The corona on the occasion of this eclipse was a very fine one, and some magnificent groups of prominences were plainly visible to the naked eye (see the Frontispiece).

The next total eclipse after that of 1905 was one which occurred on January

14, 1907. It passed across Central Asia and Siberia, and had a totality lasting two and a half minutes at most; but it was not observed as the weather was extremely bad, a circumstance not surprising with regard to those regions at that time of year.

The eclipse of January 3, 1908, passed across the Pacific Ocean. Only two small coral islands--Hull Island in the Phoenix Group, and Flint Island about 400 miles north of Tahiti--lay in the track. Two expeditions set out to observe it, i.e. a combined American party from the Lick Observatory and the Smithsonian Institution of Washington, and a private one from England under Mr. F.K. McClean. As Hull Island afforded few facilities, both parties installed their instruments on Flint Island, although it was very little better. The duration of the total phase was fairly long--about four minutes, and the sun very favourably placed, being nearly overhead. Heavy rain and clouds, however, marred observation during the first minute of totality, but the remaining three minutes were successfully utilised, good photographs of the corona being obtained.

The next few years to come are unfortunately by no means favourable from the point of view of the eclipse observer. An eclipse will take place on June 17, 1909, the track stretching from Greenland across the North Polar regions into Siberia. The geographical situation is, however, a very awkward one, and totality will be extremely short--only six seconds in Greenland and twenty-three seconds in Siberia.

The eclipse of May 9, 1910, will be visible in Tasmania. Totality will last so long as four minutes, but the sun will be at the time much too low in the sky for good observation.

The eclipse of the following year, April 28, 1911, will also be confined, roughly speaking, to the same quarter of the earth, the track passing across the old convict settlement of Norfolk Island, and then out into the Pacific.

The eclipse of April 17, 1912, will stretch from Portugal, through France and Belgium into North Germany. It will, however, be of practically no service to astronomy. Totality, for instance, will last for only three seconds in Portugal; and, though Paris lies in the central track, the eclipse, which begins as barely total, will have changed into an annular one by the time it passes over that city.

The first really favourable eclipse in the near future will be that of August 21, 1914. Its track will stretch from Greenland across Norway, Sweden, and Russia. This eclipse is a return, after one saros, of the eclipse of August 9, 1896.

The last solar eclipse which we will touch upon is that predicted for June 29, 1927. It has been already alluded to as the first of those in the future to be total in England. The central line will stretch from Wales in a north-easterly direction. Stonyhurst Observatory, in Lancashire, will lie in the track; but totality there will be very short, only about twenty seconds in duration.

[6] Knowledge, vol. xx. p. 9, January 1897.

[7] The first photographic representation of the corona had, however, been made during the eclipse of 1851. This was a daguerreotype taken by Dr. Busch at Konigsberg in Prussia.

CHAPTER IX

FAMOUS ECLIPSES OF THE MOON

The earliest lunar eclipse, of which we have any trustworthy information, was a total one which took place on the 19th March, 721 B.C., and was observed from Babylon. For our knowledge of this eclipse we are indebted to Ptolemy, the astronomer, who copied it, along with two others, from the records of the reign of the Chaldean king, Merodach-Baladan.

The next eclipse of the moon worth noting was a total one, which took place some three hundred years later, namely, in 425 B.C. This eclipse was observed at Athens, and is mentioned by Aristophanes in his play, The Clouds.

Plutarch relates that a total eclipse of the moon, which occurred in 413 B.C., so greatly frightened Nicias, the general of the Athenians, then warring in Sicily, as to cause a delay in his retreat from Syracuse which led to the destruction of his whole army.

Seven years later--namely, in 406 B.C., the twenty-sixth year of the Peloponnesian War--there took place another total lunar eclipse of which mention is made by Xenophon.

Omitting a number of other eclipses alluded to by ancient writers, we come to one recorded by Josephus as having occurred a little before the death of Herod the Great. It is probable that the eclipse in question was the total lunar one, which calculation shows to have taken place on the 15th September 5 B.C., and to have been visible in Western Asia. This is very important, for we are thus enabled to fix that year as the date of the birth of Christ, for Herod is known to have died in the early part of the year following the Nativity.

In those accounts of total lunar eclipses, which have come down to us from the Dark and Middle Ages, the colour of the moon is nearly always likened to "blood." On the other hand, in an account of the eclipse of January 23, A.D. 753, our satellite is described as "covered with a horrid black shield." We thus have examples of the two distinct appearances alluded to in Chapter VII., i.e. when the moon appears of a coppery-red colour, and when it is entirely darkened.

It appears, indeed, that, in the majority of lunar eclipses on record, the moon has appeared of a ruddy, or rather of a coppery hue, and the details on its surface have been thus rendered visible. One of the best examples of a bright eclipse of this kind is that of the 19th March 1848, when the illumination of our satellite was so great that many persons could not believe that an eclipse was actually taking place. A certain Mr. Foster, who observed this eclipse from Bruges, states that the markings on the lunar disc were almost as visible as on an "ordinary dull moonlight night." He goes on to say that the British Consul at Ghent, not knowing that there had been any eclipse, wrote to him for an explanation of the red colour of the moon on that evening.

Out of the dark eclipses recorded, perhaps the best example is that of May 18, 1761, observed by Wargentin at Stockholm. On this occasion the lunar disc is said to have disappeared so completely, that it could not be discovered even with the telescope. Another such instance is the eclipse of June 10, 1816, observed from London. The summer of that year was particularly wet--a point worthy of notice in connection with the theory that these different appearances are due to the varying state of our earth's atmosphere.

Sometimes, indeed, it has happened that an eclipse of the moon has partaken of both appearances, part of the disc being visible and part invisible. An instance of this occurred in the eclipse of July 12, 1870, when the late Rev. S.J. Johnson, one of the leading authorities on eclipses, who observed it, states that he found one-half the moon's surface quite invisible, both with the naked eye and with the telescope.

In addition to the examples given above, there are three total lunar eclipses which deserve especial mention.

1. A.D. 755, November 23. During the progress of this eclipse the moon occulted the star Aldebaran in the constellation of Taurus.

2. A.D. 1493, April 2. This is the celebrated eclipse which is said to have so well served the purposes of Christopher Columbus. Certain natives having refused to supply him with provisions when in sore straits, he announced to them that the moon would be darkened as a sign of the anger of heaven. When the event duly came to pass, the savages were so terrified that they brought him provisions as much as he needed.

3. A.D. 1610, July 6. The eclipse in question is notable as having been seen through the telescope, then a recent invention. It was without doubt the first so observed, but unfortunately the name of the observer has not come down to us.

CHAPTER X

THE GROWTH OF OBSERVATION

The earliest astronomical observations must have been made in the Dawn of Historic Time by the men who tended their flocks upon the great plains. As they watched the clear night sky they no doubt soon noticed that, with the exception of the moon and those brilliant wandering objects known to us as the planets, the individual stars in the heaven remained apparently fixed with reference to each other. These seemingly changeless points of light came in time to be regarded as sign-posts to guide the wanderer across the trackless desert, or the voyager upon the wide sea.

Just as when looking into the red coals of a fire, or when watching the clouds, our imagination conjures up strange and grotesque forms, so did the men of old see in the grouping of the stars the outlines of weird and curious shapes. Fed with mythological lore, they imagined these to be rough representations of ancient heroes and fabled beasts, whom they supposed to have been elevated to the heavens as a reward for great deeds done upon the earth. We know these groupings of stars to-day under the name of the Constellations. Looking up at them we find it extremely difficult to fit in the majority with the figures which the ancients believed them to represent. Nevertheless, astronomy has accepted the arrangement, for want of a better method of fixing the leading stars in the memory.

Our early ancestors lived the greater part of their lives in the open air, and so came to pay more attention in general to the heavenly orbs than we do. Their clock and their calendar was, so to speak, in the celestial vault. They regulated their hours, their days, and their nights by the changing positions of the sun, the moon, and the stars; and recognised the periods of seed-time and harvest, of calm and stormy weather, by the rising or setting of certain well-known constellations. Students of the classics will recall many allusions to this, especially in the Odes of Horace.

As time went on and civilisation progressed, men soon devised measuring instruments, by means of which they could note the positions of the celestial bodies in the sky with respect to each other; and, from observations thus made, they constructed charts of the stars. The earliest complete survey of this kind, of which we have a record, is the great Catalogue of stars which was made, in the second century B.C., by the celebrated Greek astronomer, Hipparchus, and in which he is said to have noted down about 1080 stars.

It is unnecessary to follow in detail the tedious progress of astronomical discovery prior to the advent of the telescope. Certain it is that, as time went on, the measuring instruments to which we have alluded had become greatly improved; but, had they even been perfect, they would have been utterly inadequate to reveal those minute displacements, from which we have learned the actual distance of the nearest of the celestial orbs. From the early times, therefore, until the medieval period of our own era, astronomy grew up upon a faulty basis, for the earth ever seemed so much the largest body in the universe, that it continued from century to century to be regarded as the very centre of things.

To the Arabians is due the credit of having kept alive the study of the stars during the dark ages of European history. They erected some fine observatories, notably in Spain and in the neighbourhood of Bagdad. Following them, some of the Oriental peoples embraced the science in earnest; Ulugh Beigh, grandson of the famous Tamerlane, founding, for instance, a great observatory at Samarcand in Central Asia. The Mongol emperors of India also established large astronomical instruments in the chief cities of their empire. When the revival of learning took place in the West, the Europeans came to the front once more in science, and rapidly forged ahead of those who had so assiduously kept alight the lamp of knowledge through the long centuries.

The dethronement of the older theories by the Copernican system, in which the earth was relegated to its true place, was fortunately soon followed by an invention of immense import, the invention of the Telescope. It is to this instrument, indeed, that we are indebted for our knowledge of the actual scale of the celestial distances. It penetrated the depths of space; it brought the distant orbs so near, that men could note the detail on the planets, or measure the small changes in their positions in the sky which resulted from the movement of our own globe.

It was in the year 1609 that the telescope was first constructed. A year or so previous to this a spectacle-maker of Middleburgh in Holland, one Hans Lippershey, had, it appears, hit upon the fact that distant objects, when viewed through certain glass lenses suitably arranged, looked nearer.[8] News of this discovery reached the ears of Galileo Galilei, of Florence, the foremost philosopher of the day, and he at once applied his great scientific attainments to the construction of an instrument based upon this principle. The result was what was called an "optick tube," which magnified distant objects some few times. It was not much larger than what we nowadays contemptuously refer to as a "spy-glass," yet its employment upon the leading celestial objects instantly sent astronomical science onward with a bound. In rapid succession Galileo announced world-moving discoveries; large spots upon the face of the sun; crater-like mountains upon the moon;

four subordinate bodies, or satellites, circling around the planet Jupiter; and a strange appearance in connection with Saturn, which later telescopic observers found to be a broad flat ring encircling that planet. And more important still, the magnified image of Venus showed itself in the telescope at certain periods in crescent and other forms; a result which Copernicus is said to have announced should of necessity follow if his system were the true one.

The discoveries made with the telescope produced, as time went on, a great alteration in the notions of men with regard to the universe at large. It must have been, indeed, a revelation to find that those points of light which they called the planets, were, after all, globes of a size comparable with the earth, and peopled perchance with sentient beings. Even to us, who have been accustomed since our early youth to such an idea, it still requires a certain stretch of imagination to enlarge, say, the Bright Star of Eve, into a body similar in size to our earth. The reader will perhaps recollect Tennyson's allusion to this in Locksley Hall, Sixty Years After:--

"Hesper--Venus--were we native to that splendour or in Mars, We should see the Globe we groan in, fairest of their evening stars.

"Could we dream of wars and carnage, craft and madness, lust and spite, Roaring London, raving Paris, in that point of peaceful light?"

The form of instrument as devised by Galileo is called the Refracting Telescope, or "Refractor." As we know it to-day it is the same in principle as his "optick tube," but it is not quite the same in construction. The early object-glass, or large glass at the end, was a single convex lens (see Fig. 8, p. 113, "Galilean"); the modern one is, on the other hand, composed of two lenses fitted together. The attempts to construct large telescopes of the Galilean type met in course of time with a great difficulty. The magnified image of the object observed was not quite pure; its edges, indeed, were fringed with rainbow-like colours. This defect was found to be aggravated with increase in the size of object-glasses. A method was, however, discovered of diminishing this colouration, or chromatic aberration as it is called from the Greek word [chra] (chroma), which means colour, viz. by making telescopes of great length and only a few inches in width. But the remedy was, in a way, worse than the disease; for telescopes thus became of such huge proportions as to be too unwieldy for use. Attempts were made to evade this unwieldiness by constructing them with skeleton tubes (see Plate II., p. 110), or, indeed, even without tubes at all; the object-glass in the tubeless or "aerial" telescope being fixed at the top of a high post, and the eye-piece, that small lens or combination of lenses, which the eye looks directly into, being kept in line with it by means of a string and manoeuvred about near the ground (Plate III., p. 112). The idea of a telescope without a tube may appear a contradiction in terms; but it is not really so, for the tube adds nothing to the magnifying power of the instrument, and is, in fact, no more than a mere device for keeping the object-glass and eye-piece in a straight line, and for preventing the observer from being hindered by stray lights in his neighbourhood. It goes without saying, of course, that the image of a celestial object will be more clear and defined when examined in the darkness of a tube.

The ancients, though they knew nothing of telescopes, had, however, found out the merit of a tube in this respect; for they employed simple tubes, blackened on the inside, in order to obtain a clearer view of distant objects. It is said that Julius Caesar, before crossing the Channel, surveyed the opposite coast of Britain through a tube of this kind.

This instrument, 150 feet in length, with a skeleton tube, was constructed by the celebrated seventeenth century astronomer, Hevelius of Danzig. From an illustration in the Machina Celestis.

A few of the most famous of the immensely long telescopes above alluded to are worthy of mention. One of these, 123 feet in length, was presented to the Royal Society of London by the Dutch astronomer Huyghens. Hevelius of Danzig constructed a skeleton one of 150 feet in length (see Plate II., p. 110). Bradley used a tubeless one 212 feet long to measure the diameter of Venus in 1722; while one of 600 feet is said to have been constructed, but to have proved quite unworkable!

Such difficulties, however, produced their natural result. They set men at work to devise another kind of telescope. In the new form, called the Reflecting Telescope, or "Reflector," the light coming from the object under observation was reflected into the eye-piece from the surface of a highly polished concave metallic mirror, or speculum, as it was called. It is to Sir

Isaac Newton that the world is indebted for the reflecting telescope in its best form. That philosopher had set himself to investigate the causes of the rainbow-like, or prismatic colours which for a long time had been such a source of annoyance to telescopic observers; and he pointed out that, as the colours were produced in the passage of the rays of light through the glass, they would be entirely absent if the light were reflected from the surface of a mirror instead.

The reflecting telescope, however, had in turn certain drawbacks of its own. A mirror, for instance, can plainly never be polished to such a high degree as to reflect as much light as a piece of transparent glass will let through. Further, the position of the eye-piece is by no means so convenient. It cannot, of course, be pointed directly towards the mirror, for the observer would then have to place his head right in the way of the light coming from the celestial object, and would thus, of course, cut it off. In order to obviate this difficulty, the following device was employed by Newton in his telescope, of which he constructed his first example in 1668. A small, flat mirror was fixed by thin wires in the centre of the tube of the telescope, and near to its open end. It was set slant-wise, so that it reflected the rays of light directly into the eye-piece, which was screwed into a hole at the side of the tube (see Fig. 8, p. 113, "Newtonian").

Although the Newtonian form of telescope had the immense advantage of doing away with the prismatic colours, yet it wasted a great deal of light; for the objection in this respect with regard to loss of light by reflection from the large mirror applied, of course, to the small mirror also. In addition, the position of the "flat," as the small mirror is called, had the further effect of excluding from the great mirror a certain proportion of light. But the reflector had the advantage, on the other hand, of costing less to make than the refractor, as it was not necessary to procure flawless glass for the purpose. A disc of a certain metallic composition, an alloy of copper and tin, known in consequence as speculum metal, had merely to be cast; and this had to be ground and polished upon one side only, whereas a lens has to be thus treated upon both its sides. It was, therefore, possible to make a much larger instrument at a great deal less labour and expense.

We have given the Newtonian form as an example of the principle of the reflecting telescope. A somewhat similar instrument had, however, been projected, though not actually constructed, by James Gregory a few years earlier than Newton's, i.e. in 1663. In this form of reflector, known as the "Gregorian" telescope, a hole was made in the big concave mirror; and a small mirror, also concave, which faced it at a certain distance, received the reflected rays, and reflected them back again through the hole in question into the eye-piece, which was fixed just behind (see Fig. 8, p. 113, "Gregorian"). The Gregorian had thus the sentimental advantage of being pointed directly at the object. The hole in the big mirror did not cause any loss of light, for the central portion in which it was made was anyway unable to receive light through the small mirror being directly in front of it. An adaptation of the Gregorian was the "Cassegrainian" telescope, devised by Cassegrain in 1672, which differed from it chiefly in the small mirror being convex instead of concave (see Fig. 8, p. 113, "Cassegrainian"). These direct-view forms of the reflecting telescope were much in vogue about the middle of the eighteenth century, when many beautiful examples of Gregorians were made by the famous optician, James Short, of Edinburgh.

An adaptation of the Newtonian type of telescope is known as the "Herschelian," from being the kind favoured by Sir William Herschel. It is, however, only suitable in immense instruments, such as Herschel was in the habit of employing. In this form the object-glass is set at a slight slant, so that the light coming from the object is reflected straight into the eye-piece, which is fixed facing it in the side of the tube (see Fig. 8, p. 113, "Herschelian"). This telescope has an advantage over the other forms of reflector through the saving of light consequent on doing away with the second reflection. There is, however, the objection that the slant of the object-glass is productive of some distortion in the appearance of the object observed; but this slant is of necessity slight when the length of the telescope is very great.

The principle of this type of telescope had been described to the French Academy of Sciences as early as 1728 by Le Maire, but no one availed himself of the idea until 1776, when Herschel tried it. At first, however, he rejected it; but in 1786 he seems to have found that it suited the huge instruments which he was then making. Herschel's largest telescope, constructed in 1789, was about four feet in diameter and forty feet in length. It is generally spoken of as the "Forty-foot Telescope," though all other instruments have been known by their diameters, rather than by their lengths.

To return to the refracting telescope. A solution of the colour difficulty was arrived at in 1729 (two years after Newton's death) by an Essex gentleman named Chester Moor Hall. He discovered that by making a double object-glass, composed of an outer convex lens and an inner concave lens, made respectively of different kinds of glass, i.e. crown glass and flint glass, the troublesome colour effects could be, to a very great extent, removed. Hall's investigations appear to have been rather of an academic nature; and, although he is believed to have constructed a small telescope upon these lines, yet he seems to have kept the matter so much to himself that it was not until the year 1758 that the first example of the new instrument was given to the world. This was done by John Dollond, founder of the well-known optical firm of Dollond, of Ludgate Hill, London, who had, quite independently, re-discovered the principle.

This "Achromatic" telescope, or telescope "free from colour effects," is the kind ordinarily in use at present, whether for astronomical or for terrestrial purposes (see Fig. 8, p. 113, "Achromatic"). The expense of making large instruments of this type is very great, for, in the object-glass alone, no less than four surfaces have to be ground and polished to the required curves; and, usually, the two lenses of which it is composed have to fit quite close together.

With the object of evading the expense referred to, and of securing complete freedom from colour effects, telescopes have even been made, the object-glasses of which were composed of various transparent liquids placed between thin lenses; but leakages, and currents set up within them by changes of temperature, have defeated the ingenuity of those who devised these substitutes.

The solution of the colour difficulty by means of Dollond's achromatic refractor has not, however, ousted the reflecting telescope in its best, or Newtonian form, for which great concave mirrors made of glass, covered with a thin coating of silver and highly polished, have been used since about 1870 instead of metal mirrors. They are very much lighter in weight and cheaper to make than the old specula; and though the silvering, needless to say, deteriorates with time, it can be renewed at a comparatively trifling cost. Also these mirrors reflect much more light, and give a clearer view, than did the old metallic ones.

When an object is viewed through the type of astronomical telescope ordinarily in use, it is seen upside down. This is, however, a matter of very small moment in dealing with celestial objects; for, as they are usually round, it is really not of much consequence which part we regard as top and which as bottom. Such an inversion would, of course, be most inconvenient when viewing terrestrial objects. In order to observe the latter we therefore employ what is called a terrestrial telescope, which is merely a refractor with some extra lenses added in the eye portion for the purpose of turning the inverted image the right way up again. These extra lenses, needless to say, absorb a certain amount of light; wherefore it is better in astronomical observation to save light by doing away with them, and putting up with the slight inconvenience of seeing the object inverted.

This inversion of images by the astronomical telescope must be specially borne in mind with regard to the photographs of the moon in Chapter XVI.

In the year 1825 the largest achromatic refractor in existence was one of nine and a half inches in diameter constructed by Fraunhofer for the Observatory of Dorpat in Russia. The largest refractors in the world to-day are in the United States, i.e. the forty-inch of the Yerkes Observatory (see Plate IV., p. 118), and the thirty-six inch of the Lick. The object-glasses of these and of the thirty-inch telescope of the Observatory of Pulkowa, in Russia, were made by the great optical house of Alvan Clark & Sons, of Cambridge, Massachusetts, U.S.A. The tubes and other portions of the Yerkes and Lick telescopes were, however, constructed by the Warner and Swasey Co., of Cleveland, Ohio.

The largest reflector, and so the largest telescope in the world, is still the six-foot erected by the late Lord Rosse at Parsonstown in Ireland, and completed in the year 1845. It is about fifty-six feet in length. Next come two of five feet, with mirrors of silver on glass; one of them made by the late Dr. Common, of Ealing, and the other by the American astronomer, Professor G.W. Ritchey. The latter of these is installed in the Solar Observatory belonging to Carnegie Institution of Washington, which is situated on Mount Wilson in California. The former is now at the Harvard College Observatory, and is considered by Professor Moulton to be probably the most efficient reflector in use at present. Another large reflector is the three-foot made by Dr. Common. It came into the possession of Mr. Crossley of Halifax, who presented it to the Lick Observatory, where it is now known as the "Crossley Reflector."

Although to the house of Clark belongs, as we have seen, the credit of constructing the object-glasses of the largest refracting telescopes of our time, it has nevertheless keen competitors in Sir Howard Grubb, of Dublin, and such well-known firms as Cooke of York and Steinheil of Munich. In the four-foot reflector, made in 1870 for the Observatory of Melbourne by the firm of Grubb, the Cassegrainian principle was employed.

With regard to the various merits of refractors and reflectors much might be said. Each kind of instrument has, indeed, its special advantages; though perhaps, on the whole, the most perfect type of telescope is the achromatic refractor.

Great telescope at the Yerkes Observatory of the University of Chicago, Williams Bay, Wisconsin, U.S.A. It was erected in 1896-7, and is the largest refracting telescope in the world. Diameter of object-glass, 40 inches; length of telescope, about 60 feet. The object-glass was made by the firm of Alvan Clark and Sons, of Cambridge, Massachusetts; the other portions of the instrument by the Warner and Swasey Co., of Cleveland, Ohio.

In connection with telescopes certain devices have from time to time been introduced, but these merely aim at the convenience of the observer and do not supplant the broad principles upon which are based the various types of instrument above described. Such, for instance, are the "Siderostat," and another form of it called the "Coelostat," in which a plane mirror is made to revolve in a certain manner, so as to reflect those portions of the sky which are to be observed, into the tube of a telescope kept fixed. Such too are the "Equatorial Coud? of the late M. Loewy, Director of the Paris Observatory, and the "Sheepshanks Telescope" of the Observatory of Cambridge, in which a telescope is separated into two portions, the eye-piece portion being fixed upon a downward slant, and the object-glass portion jointed to it at an angle and pointed up at the sky. In these two instruments (which, by the way, differ materially) an arrangement of slanting mirrors in the tubes directs the journey of the rays of light from the object-glass to the eye-piece. The observer can thus sit at the eye-end of his telescope in the warmth and comfort of his room, and observe the stars in the same unconstrained manner as if he were merely looking down into a microscope.

Needless to say, devices such as these are subject to the drawback that the mirrors employed sap a certain proportion of the rays of light. It will be remembered that we made allusion to loss of light in this way, when pointing out the advantage in light grasp of the Herschelian form of telescope, where only one reflection takes place, over the Newtonian in which there are two.

It is an interesting question as to whether telescopes can be made much larger. The American astronomer, Professor G.E. Hale, concludes that the limit of refractors is about five feet in diameter, but he thinks that reflectors as large as nine feet in diameter might now be made. As regards refractors there are several strong reasons against augmenting their proportions. First of all comes the great cost. Secondly, since the lenses are held in position merely round their rims, they will bend by their weight in the centres if they are made much larger. On the other hand, attempts to obviate this, by making the lenses thicker, would cause a decrease in the amount of light let through.

But perhaps the greatest stumbling-block to the construction of larger telescopes is the fact that the unsteadiness of the air will be increasingly magnified. And further, the larger the tubes become, the more difficult will it be to keep the air within them at one constant temperature throughout their lengths.

It would, indeed, seem as if telescopes are not destined greatly to increase in size, but that the means of observation will break out in some new direction, as it has already done in the case of photography and the spectroscope. The direct use of the eye is gradually giving place to indirect methods. We are, in fact, now feeling rather than seeing our way about the universe. Up to the present, for instance, we have not the slightest proof that life exists elsewhere than upon our earth. But who shall say that the twentieth century has not that in store for us, by which the presence of life in other orbs may be perceived through some form of vibration transmitted across illimitable space? There is no use speaking of the impossible or the inconceivable. After the extraordinary revelations of the spectroscope--nay, after the astounding discovery of Roentgen--the word impossible should be cast aside, and inconceivability cease to be regarded as any criterion.

[8] The principle upon which the telescope is based appears to have been known theoretically for a long time previous to this. The monk Roger Bacon, who lived in the thirteenth century, describes it very clearly; and several writers of the sixteenth century have also dealt with the idea. Even Lippershey's claims to a practical solution of the question were hotly contested at the time by two of his own countrymen, i.e. a certain Jacob Metius, and another spectacle-maker of Middleburgh, named Jansen.

CHAPTER XI

SPECTRUM ANALYSIS

If white light (that of the sun, for instance) be passed through a glass prism, namely, a piece of glass of triangular shape, it will issue from it in rainbow-tinted colours. It is a common experience with any of us to notice this when the sunlight shines through cut-glass, as in the pendant of a chandelier, or in the stopper of a wine-decanter.

The same effect may be produced when light passes through water. The Rainbow, which we all know so well, is merely the result of the sunlight passing through drops of falling rain.

White light is composed of rays of various colours. Red, orange, yellow, green, blue, indigo, and violet, taken all together, go, in fact, to make up that effect which we call white.

It is in the course of the refraction, or bending of a beam of light, when it passes in certain conditions through a transparent and denser medium, such as glass or water, that the constituent rays are sorted out and spread in a row according to their various colours. This production of colour takes place usually near the edges of a lens; and, as will be recollected, proved very obnoxious to the users of the old form of refracting telescope.

It is, indeed, a strange irony of fate that this very same production of colour, which so hindered astronomy in the past, should have aided it in recent years to a remarkable degree. If sunlight, for instance, be admitted through a narrow slit before it falls upon a glass prism, it will issue from the latter in the form of a band of variegated colour, each colour blending insensibly with the next. The colours arrange themselves always in the order which we have mentioned. This seeming band is, in reality, an array of countless coloured images of the original slit ranged side by side; the colour of each image being the slightest possible shade different from that next to it. This strip of colour when produced by sunlight is called the "Solar Spectrum" (see Fig. 9, p. 123). A similar strip, or spectrum, will be produced by any other light; but the appearance of the strip, with regard to preponderance of particular colours, will depend upon the character of that light. Electric light and gas light yield spectra not unlike that of sunlight; but that of gas is less rich in blue and violet than that of the sun.

The Spectroscope, an instrument devised for the examination of spectra, is, in its simplest form, composed of a small tube with a narrow slit and prism at one end, and an eye-piece at the other. If we drop ordinary table salt into the flame of a gas light, the flame becomes strongly yellow. If, then, we observe this yellow flame with the spectroscope, we find that its spectrum consists almost entirely of two bright yellow transverse lines. Chemically considered ordinary table salt is sodium chloride; that is to say, a compound of the metal sodium and the gas chlorine. Now if other compounds of sodium be experimented with in the same manner, it will soon be found that these two yellow lines are characteristic of sodium when turned into vapour by great heat. In the same manner it can be ascertained that every element, when heated to a condition of vapour, gives as its spectrum a set of lines peculiar to itself. Thus the spectroscope enables us to find out the composition of substances when they are reduced to vapour in the laboratory.

In order to increase the power of a spectroscope, it is necessary to add to the number of prisms. Each extra prism has the effect of lengthening the coloured strip still more, so that lines, which at first appeared to be single merely through being crowded together, are eventually drawn apart and become separately distinguishable.

On this principle it has gradually been determined that the sun is composed of elements similar to those which go to make up our earth. Further, the composition of the stars can be ascertained in the same manner; and we find them formed on a like pattern, though with certain elements in greater or less proportion as the case may be. It is in consequence of our thus definitely ascertaining that the stars are self-luminous, and of a sun-like character, that we are enabled to speak of them as suns, or to call the sun a star.

In endeavouring to discover the elements of which the planets and satellites of our system are composed, we, however, find ourselves baffled, for the simple reason that these bodies emit no real light of their own. The light which reaches us from them, being merely reflected sunlight, gives only the ordinary solar spectrum when examined with the spectroscope. But in certain cases we find that the solar spectrum thus viewed shows traces of being weakened, or rather of suffering absorption; and it is concluded that this may be due to the sunlight having had to pass through an atmosphere on its way to and from the surface of the planet from which it is reflected to us.

Since the sun is found to be composed of elements similar to those which go to make up our earth, we need not be disheartened at this failure of the spectroscope to inform us of the composition of the planets and satellites. We are justified, indeed, in assuming that more or less the same constituents run through our solar system; and that the elements of which these bodies are composed are similar to those which are found upon our earth and in the sun.

The spectroscope supplies us with even more information. It tells us, indeed, whether the sun-like body which we are observing is moving away from us or towards us. A certain slight shifting of the lines towards the red or violet end of the spectrum respectively, is found to follow such movement. This method of observation is known by the name of Doppler's Method,[9] and by it we are enabled to confirm the evidence which the sunspots give us of the rotation of the sun; for we find thus that one edge of that body is continually approaching us, and the other edge is continually receding from us. Also, we can ascertain in the same manner that certain of the stars are moving towards us, and certain of them away from us.

[9] The idea, initiated by Christian Doppler at Prague in 1842, was originally applied to sound. The approach or recession of a source from which sound is coming is invariably accompanied by alterations of pitch, as the reader has no doubt noticed when a whistling railway-engine has approached him or receded from him. It is to Sir William Huggins, however, that we are indebted

for the application of the principle to spectroscopy. This he gave experimental proof of in the year 1868.

CHAPTER XII

THE SUN

The sun is the chief member of our system. It controls the motions of the planets by its immense gravitative power. Besides this it is the most important body in the entire universe, so far as we are concerned; for it pours out continually that flood of light and heat, without which life, as we know it, would quickly become extinct upon our globe.

Light and heat, though not precisely the same thing, may be regarded, however, as next-door neighbours. The light rays are those which directly affect the eye and are comprised in the visible spectrum. We feel the heat rays, the chief of which are beyond the red portion of the spectrum. They may be investigated with the bolometer, an instrument invented by the late Professor Langley. Chemical rays--for instance, those radiations which affect the photographic plate--are for the most part also outside the visible spectrum. They are, however, at the other end of it, namely, beyond the violet.

Such a scale of radiations may be compared to the keyboard of an imaginary piano, the sound from only one of whose octaves is audible to us.

The brightest light we know on the earth is dull compared with the light of the sun. It would, indeed, look quite dark if held up against it.

It is extremely difficult to arrive at a precise notion of the temperature of the body of the sun. However, it is far in excess of any temperature which we can obtain here, even in the most powerful electric furnace.

A rough idea of the solar heat may be gathered from the calculation that if the sun's surface were coated all over with a layer of ice 4000 feet thick, it would melt through this completely in one hour.

The sun cannot be a hot body merely cooling; for the rate at which it is at present giving off heat could not in such circumstances be kept up, according to Professor Moulton, for more than 3000 years. Further, it is not a mere burning mass, like a coal fire, for instance; as in that case about a thousand years would show a certain drop in temperature. No perceptible diminution of solar heat having taken place within historic experience, so far as can be ascertained, we are driven to seek some more abstruse explanation.

The theory which seems to have received most acceptance is that put forward by Helmholtz in 1854. His idea was that gravitation produces continual contraction, or falling in of the outer parts of the sun; and that this falling in, in its turn, generates enough heat to compensate for what is being given off. The calculations of Helmholtz showed that a contraction of about 100 feet a year from the surface towards the centre would suffice for the purpose. In recent years, however, this estimate has been extended to about 180 feet. Nevertheless, even with this increased figure, the shrinkage required is so slight in comparison with the immense girth of the sun, that it would take a continual contraction at this rate for about 6000 years, to show even in our finest telescopes that any change in the size of that body was taking place at all. Upon this assumption of continuous contraction, a time should, however, eventually be reached when the sun will have shrunk to such a degree of solidity, that it will not be able to shrink any further. Then, the loss of heat not being made up for any longer, the body of the sun should begin to grow cold. But we need not be distressed on this account; for it will take some 10,000,000 years, according to the above theory, before the solar orb becomes too cold to support life upon our earth.

Since the discovery of radium it has, on the other hand, been suggested, and not unreasonably, that radio-active matter may possibly play an important part in keeping up the heat of the sun. But the body of scientific opinion appears to consider the theory of contraction as a result of gravitation, which has been outlined above, to be of itself quite a sound explanation. Indeed, the late Lord Kelvin is said to have held to the last that it was amply sufficient to account for the underground heat of the earth, the heat of the sun, and that of all the stars in the universe.

One great difficulty in forming theories with regard to the sun, is the fact that the temperature and gravitation there are enormously in excess of anything we meet with upon our earth. The force of gravity at the sun's surface is, indeed, about twenty-seven times that at the surface of our globe.

The earth's atmosphere appears to absorb about one-half of the radiations which come to us from the sun. This absorptive effect is very noticeable when the solar orb is low down in our sky, for its light and heat are then clearly much reduced. Of the light rays, the blue ones are the most easily absorbed in this way; which explains why the sun looks red when near the horizon. It has then, of course, to shine through a much greater thickness of atmosphere than when high up in the heavens.

What astonishes one most about the solar radiation, is the immense amount of it that is apparently wasted into space in comparison with what falls directly upon the bodies of the solar system. Only about the one-hundred-millionth is caught by all the planets together. What becomes of the rest we cannot tell.

That brilliant white body of the sun, which we see, is enveloped by several layers of gases and vaporous matter, in the same manner as our globe is enveloped by its atmosphere (see Fig. 10, p. 131). These are transparent, just as our atmosphere is transparent; and so we see the white bright body of the sun right through them.

This white bright portion is called the Photosphere. From it comes most of that light and heat which we see and feel. We do not know what lies under the photosphere, but, no doubt, the more solid portions of the sun are there situated. Just above the photosphere, and lying close upon it, is a veil of smoke-like haze.

Next upon this is what is known as the Reversing Layer, which is between 500 and 1000 miles in thickness. It is cooler than the underlying photosphere, and is composed of glowing gases. Many of the elements which go to make up our earth are present in the reversing layer in the form of vapour.

The Chromosphere, of which especial mention has already been made in dealing with eclipses of the sun, is another layer lying immediately upon the last one. It is between 5000 and 10,000 miles in thickness. Like the reversing layer, it is composed of glowing gases, chief among which is the vapour of hydrogen. The colour of the chromosphere is, in reality, a brilliant scarlet; but, as we have already said, the intensely white light of the photosphere shines through it from behind, and entirely overpowers its redness. The upper portion of the chromosphere is in violent agitation, like the waves of a stormy sea, and from it rise those red prominences which, it will be recollected, are such a notable feature in total solar eclipses.

The Corona lies next in order outside the chromosphere, and is, so far as we know, the outermost of the accompaniments of the sun. This halo of pearly-white light is irregular in outline, and fades away into the surrounding sky. It extends outwards from the sun to several millions of miles. As has been stated, we can never see the corona unless, when during a total solar eclipse, the moon has, for the time being, hidden the brilliant photosphere completely from our view.

The solar spectrum is really composed of three separate spectra commingled, i.e. those of the photosphere, of the reversing layer, and of the chromosphere respectively.

If, therefore, the photosphere could be entirely removed, or covered up, we should see only the spectra of those layers which lie upon it. Such a state of things actually occurs in a total eclipse of the sun. When the moon's body has crept across the solar disc, and hidden the last piece of photosphere, the solar spectrum suddenly becomes what is technically called "reversed,"--the dark lines crossing it changing into bright lines. This occurs because a strip of those layers which lie immediately upon the photosphere remains still uncovered. The lower of these layers has therefore been called the "reversing layer," for want of a better name. After a second or two this reversed spectrum mostly vanishes, and an altered spectrum is left to view. Taking into consideration the rate at which the moon is moving across the face of the sun, and the very short time during which the spectrum of the reversing layer lasts, the thickness of that layer is estimated to be not more than a few hundred miles. In the same way the last of the three spectra--namely, that of the chromosphere--remains visible for such a time as allows us to estimate its depth at about ten times that of the reversing layer, or several thousand miles.

When the chromosphere, in its turn during a total eclipse, has been covered by the moon, the corona alone is left. This has a distinct spectrum of its own also; wherein is seen a strange line in the green portion, which does not tally with that of any element we are acquainted with upon the earth. This unknown element has received for the time being the name of "Coronium."

CHAPTER XIII

THE SUN--continued

The various parts of the Sun will now be treated of in detail.

I. PHOTOSPHERE.

The photosphere, or "light-sphere," from the Greek [ph] (phos), which means light, is, as we have already said, the innermost portion of the sun which can be seen. Examined through a good telescope it shows a finely mottled structure, as of brilliant granules, somewhat like rice grains, with small dark spaces lying in between them. It has been supposed that we have here the process of some system of circulation by which the sun keeps sending forth its radiations. In the bright granules we perhaps see masses of intensely heated matter, rising from the interior of the sun. The dark interspaces may represent matter which has become cooled and darkened through having parted with its heat and light, and is falling back again into the solar furnace.

The sun spots, so familiar to every one nowadays, are dark patches which are often seen to break out in the photosphere (see Plate V., p. 134). They last during various periods of time; sometimes only for a few days, sometimes so long as a month or more. A spot is usually composed of a dark central portion called the umbra, and a less dark fringe around this called the penumbra (see Plate VI., p. 136). The umbra ordinarily has the appearance of a deep hole in the photosphere; but, that it is a hole at all, has by no means been definitely proved.

Sun spots are, as a rule, some thousands of miles across. The umbra of a good-sized spot could indeed engulf at once many bodies the size of our earth.

Sun spots do not usually appear singly, but in groups. The total area of a group of this kind may be of immense extent; even so great as to cover the one-hundredth part of the whole surface of the sun. Very large spots, when such are present, may be seen without any telescope; either through a piece of smoked glass, or merely with the naked eye when the air is misty, or the sun low on the horizon.

The umbra of a spot is not actually dark. It only appears so in contrast with the brilliant photosphere around.

Spots form, grow to a large size in comparatively short periods of time, and then quickly disappear. They seem to shrink away as a consequence of the photosphere closing in upon them.

That the sun is rotating upon an axis, is shown by the continual change of position of all spots in one constant direction across his disc. The time in which a spot is carried completely round depends, however, upon the position which it occupies upon the sun's surface. A spot situated near the equator of the sun goes round once in about twenty-five days. The further a spot is situated from this equator, the longer it takes. About twenty-seven days is the time taken by a spot situated midway between the equator and the solar poles. Spots occur to the north of the sun's equator, as well as to the south; though, since regular observations have been made--that is to say, during the past fifty years or so--they appear to have broken out a little more frequently in the southern parts.

From these considerations it will be seen that the sun does not rotate as the earth does, but that different portions appear to move at different speeds. Whether in the neighbourhood of the solar poles the time of rotation exceeds twenty-seven days we are unable to ascertain, for spots are not seen in those regions. No explanation has yet been given of this peculiar rotation; and the most we can say on the subject is that the sun is not by any means a solid body.

Facul?(Latin, little torches) are brilliant patches which appear here and there upon the sun's surface, and are in some way associated with spots. Their displacement, too, across the solar face confirms the evidence which the spots give us of the sun's rotation.

Our proofs of this rotation are still further strengthened by the Doppler spectroscopic method of observation alluded to in Chapter XI. As was then stated, one edge of the sun is thus found to be continually approaching us, and the other side continually receding from us. The varying rates of rotation, which the spots and facul?give us, are duly confirmed by this method.

The first attempt to bring some regularity into the question of sunspots was the discovery by Schwabe, in 1852, that they were subject to a regular variation. As a matter of fact they wax and wane in their number, and the total area which they cover, in the course of a period, or cycle, of on an average about 11-1/4 years; being at one part of this period large and abundant, and at another few and small. This period of 11-1/4 years is known as the sun spot cycle. No explanation has yet been given of the curious round of change, but the period in question seems to govern most of the phenomena connected with the sun.

II. REVERSING LAYER.

This is a layer of relatively cool gases lying immediately upon the photosphere. We never see it directly; and the only proof we have of its presence is that remarkable reversal of the spectrum already described, when during an instant or two in a total eclipse, the advancing edge of the moon, having just hidden the brilliant photosphere, is moving across the fine strip which the layer then presents edgewise towards us. The fleeting moments during which this reversed spectrum lasts, informs us that the layer is comparatively shallow; little more indeed than about 500 miles in depth.

The spectrum of the reversing layer, or "flash spectrum," as it is sometimes called on account of the instantaneous character with which the change takes place, was, as we have seen, first noticed by Young in 1870; and has been successfully photographed since then during several eclipses. The layer itself appears to be in a fairly quiescent state; a marked contrast to the seething photosphere beneath, and the agitated chromosphere above.

III. THE CHROMOSPHERE.

The Chromosphere--so called from the Greek [chra] (chroma), which signifies colour--is a layer of gases lying immediately upon the preceding one.

Its thickness is, however, plainly much the greater of the two; for whereas the reversing layer is only revealed to us indirectly by the spectroscope, a portion of the chromosphere may clearly be seen in a total eclipse in the form of a strip of scarlet light. The time which the moon's edge takes to traverse it tells us that it must be about ten times as deep as the reversing layer, namely, from 5000 to 10,000 miles in depth. Its spectrum shows that it is composed chiefly of hydrogen, calcium and helium, in the state of vapour. Its red colour is mainly due to glowing hydrogen. The element helium, which it also contains, has received its appellation from [h 阮 ios] (helios), the Greek name for the sun; because, at the time when it first attracted attention, there appeared to be no element corresponding to it upon our earth, and it was consequently imagined to be confined to the sun alone. Sir William Ramsay, however, discovered it to be also a terrestrial element in 1895, and since then it has come into much prominence as one of the products given off by radium.

Taking into consideration the excessive force of gravity on the sun, one would expect to find the chromosphere and reversing layer growing gradually thicker in the direction of the photosphere. This, however, is not the case. Both these layers are strangely enough of the same densities all through; which makes it suspected that, in these regions, the force of gravity may be counteracted by some other force or forces, exerting a powerful pressure outwards from the sun.

IV. THE PROMINENCES.

We have already seen, in dealing with total eclipses, that the exterior surface of the chromosphere is agitated like a stormy sea, and from it billows of flame are tossed up to gigantic heights. These flaming jets are known under the name of prominences, because they were first noticed in the form of brilliant points projecting from behind the rim of the moon when the sun was totally eclipsed. Prominences are of two kinds, eruptive and quiescent. The eruptive prominences spurt up directly from the chromosphere with immense speeds, and change their shape with great rapidity. Quiescent prominences, on the other hand, have a form somewhat like trees, and alter their shape but slowly. In the eruptive prominences glowing masses of gas are shot up to altitudes sometimes as high as 300,000 miles,[10] with velocities even so great as from 500 to 600 miles a second. It has been noticed that the eruptive prominences are mostly found in those portions of the sun where spots usually appear, namely, in the regions near the solar equator. The quiescent prominences, on the other hand, are confined, as a rule, to the neighbourhood of the sun's poles.

Prominences were at first never visible except during total eclipses of the sun. But in the year 1868, as we have already seen, a method of employing the spectroscope was devised, by means of which they could be observed and studied at any time, without the necessity of waiting for an eclipse.

A still further development of the spectroscope, the Spectroheliograph, an instrument invented almost simultaneously by Professor Hale and the French astronomer, M. Deslandres, permits of photographs being taken of the sun, with the light emanating from only one of its glowing gases at a time. For instance, we can thus obtain a record of what the glowing hydrogen alone is doing on the solar body at any particular moment. With this instrument it is also possible to obtain a series of photographs, showing what is taking place upon the sun at various levels. This is very useful in connection with the study of the spots; for we are, in consequence, enabled to gather more evidence on the subject of their actual form than is given us by their highly foreshortened appearances when observed directly in the telescope.

V. CORONA. (Latin, a Crown.)

This marvellous halo of pearly-white light, which displays itself to our view only during the total phase of an eclipse of the sun, is by no means a layer like those other envelopments of the sun of which we have just been treating. It appears, on the other hand, to be composed of filmy matter, radiating outwards in every direction, and fading away gradually into space. Its structure is noted to bear a strong resemblance to the tails of comets, or the streamers of the aurora borealis.

Our knowledge concerning the corona has, however, advanced very slowly. We have not, so far, been as fortunate with regard to it as with regard to the prominences; and, for all we can gather concerning it, we are still entirely dependent upon the changes and chances of total solar eclipses. All attempts, in fact, to apply the spectroscopic method, so as to observe the corona at leisure in full sunlight in the way in which the prominences can be observed, have up to the present met with failure.

The general form under which the corona appears to our eyes varies markedly at different eclipses. Sometimes its streamers are many, and radiate all round; at other times they are confined only to the middle portions of the sun, and are very elongated, with short feathery-looking wisps adorning the solar poles. It is noticed that this change of shape varies in close accordance with that 11-1/4 year period during which the sun spots wax and wane; the many-streamered regular type corresponding to the time of great sunspot activity, while the irregular type with the long streamers is present only when the spots are few (see Plate VII., p. 142). Streamers have often been noted to issue from those regions of the sun where active prominences are at the moment in existence; but it cannot be laid down that this is always the case.

No hypothesis has yet been formulated which will account for the structure of the corona, or for its variation in shape. The great difficulty with regard to theorising upon this subject, is the fact that we see so much of the corona under conditions of marked foreshortening. Assuming, what indeed seems natural, that the rays of which it is composed issue in every direction from the solar body, in a manner which may be roughly imitated by sticking pins all over a ball; it is plainly impossible to form any definite idea concerning streamers, which actually may owe most of the shape they present to us, to the mixing up of multitudes of rays at all kinds of angles to the line of sight. In a word, we have to try and form an opinion concerning an arrangement which, broadly speaking, is spherical, but which, on account of its distance, must needs appear to us as absolutely flat.

The most known about the composition of the corona is that it is made up of particles of matter, mingled with a glowing gas. It is an element in the composition of this gas which, as has been stated, is not found to tally with any known terrestrial element, and has, therefore, received the name of coronium for want of a better designation.

One definite conclusion appears to be reached with regard to the corona, i.e. that the matter of which it is composed, must be exceedingly rarefied; as it is not found, for instance, to retard appreciably the speed of comets, on occasions when these bodies pass very close to the sun. A calculation has indeed been made which would tend to show that the particles composing the coronal matter, are separated from each other by a distance of perhaps between two and three yards! The density of the corona is found not to increase inwards towards the sun. This is what has already been noted with regard to the layers lying beneath it. Powerful forces, acting in opposition to gravity, must hold sway here also.

The 11-1/4 year period, during which the sun spots vary in number and size, appears to govern the activities of the sun much in the same way that our year does the changing seasonal conditions of our earth. Not only, as we have seen, does the corona vary its shape in accordance with the said period, but the activity of the prominences, and of the facul? follow suit. Further, this constant round of ebb and flow is not confined to the sun itself, but, strangely enough, affects the earth also. The displays of the aurora borealis, which we experience here, coincide closely with it, as does also the varying state of the earth's magnetism. The connection may be still better appreciated when a great spot, or group of spots, has made its appearance upon the sun. It has, for example, often been noted that when the solar rotation carries a spot, or group of spots, across the middle of the visible surface of the sun, our magnetic and electrical arrangements are disturbed for the time being. The magnetic needles in our observatories are, for instance, seen to oscillate violently, telegraphic communication is for a while upset, and magnificent displays of the aurora borealis illumine our night skies. Mr. E.W. Maunder, of Greenwich Observatory, who has made a very careful investigation of this subject, suspects that, when elongated coronal streamers are whirled round in our direction by the solar rotation, powerful magnetic impulses may be projected upon us at the moments when such streamers are pointing towards the earth.

Some interesting investigations with regard to sunspots have recently been published by Mrs. E.W. Maunder. In an able paper, communicated to the Royal Astronomical Society on May 10, 1907, she reviews the Greenwich Observatory statistics dealing with the number and extent of the spots which have appeared during the period from 1889 to 1901--a whole sunspot cycle. From a detailed study of the dates in question, she finds that the number of those spots which are formed on the side of the sun turned away from us, and die out upon the side turned towards us, is much greater than the number of those which are formed on the side turned towards us and die out upon the side turned away. It used, for instance, to be considered that the influence of a planet might produce sunspots; but these investigations make it look rather as if some influence on the part of the earth tends, on the contrary, to extinguish them. Mrs. Maunder, so far, prefers to call the influence thus traced an apparent influence only, for, as she very fairly points out, it seems difficult to attribute a real influence in this matter to the earth, which is so small a thing in comparison not only with the sun, but even with many individual spots.

The above investigation was to a certain degree anticipated by Mr. Henry Corder in 1895; but Mrs. Maunder's researches cover a much longer period, and the conclusions deduced are of a wider and more defined nature.

With regard to its chemical composition, the spectroscope shows us that thirty-nine of the elements which are found upon our earth are also to be found in the sun. Of these the best known are hydrogen, oxygen, helium, carbon, calcium, aluminium, iron, copper, zinc, silver, tin, and lead. Some elements of the metallic order have, however, not been found there, as, for instance, gold and mercury; while a few of the other class of element, such as nitrogen, chlorine, and sulphur, are also absent. It must not, indeed, be concluded that the elements apparently missing do not exist at all in the solar body. Gold and mercury have, in consequence of their great atomic weight, perhaps sunk away into the centre. Again, the fact that we cannot find traces of certain other elements, is no real proof of their entire absence. Some of them may, for instance, be resolved into even simpler forms, under the unusual conditions which exist in the sun; and so we are unable to trace them with the spectroscope, the experience of which rests on laboratory experiments conducted, at best, in conditions which obtain upon the earth.

[10] On November 15, 1907, Dr. A. Rambaut, Radcliffe Observer at Oxford University, noted a prominence which rose to a height of 324,600 miles.

CHAPTER XIV

THE INFERIOR PLANETS

Starting from the centre of the solar system, the first body we meet with is the planet Mercury. It circulates at an average distance from the sun of about thirty-six millions of miles. The next body to it is the planet Venus, at about sixty-seven millions of miles, namely, about double the distance of Mercury from the sun. Since our earth comes next again, astronomers call those planets which circulate within its orbit, i.e. Mercury and Venus, the Inferior Planets, while those which circulate outside it they call the Superior Planets.[11]

In studying the inferior planets, the circumstances in which we make our observations are so very similar with regard to each, that it is best to take them together. Let us begin by considering the various positions of an inferior planet, as seen from the earth, during the course of its journeys round the sun. When furthest from us it is at the other side of the sun, and cannot then be seen owing to the blaze of light. As it continues its journey it passes to the left of the sun, and is then sufficiently away from the glare to be plainly seen. It next draws in again towards the sun, and is once more lost to view in the blaze at the time of its passing nearest to us. Then it gradually comes out to view on the right hand, separates from the sun up to a certain distance as before, and again recedes beyond the sun, and is for the time being once more lost to view.

To these various positions technical names are given. When the inferior planet is on the far side of the sun from us, it is said to be in Superior Conjunction. When it has drawn as far as it can to the left hand, and is then as east as possible of the sun, it is said to be at its Greatest Eastern Elongation. Again, when it is passing nearest to us, it is said to be in Inferior Conjunction; and, finally, when it has drawn as far as it can to the right hand, it is spoken of as being at its Greatest Western Elongation (see Fig. 11, p. 148).

The continual variation in the distance of an interior planet from us, during its revolution around the sun, will of course be productive of great alterations in its apparent size. At superior conjunction it ought, being then farthest away, to show the smallest disc; while at inferior conjunction, being the nearest, it should look much larger. When at greatest elongation, whether eastern or western, it should naturally present an appearance midway in size between the two.

From the above considerations one would be inclined to assume that the best time for studying the surface of an interior planet with the telescope is when it is at inferior conjunction, or, nearest to us. But that this is not the case will at once appear if we consider that the sunlight is then falling upon the side away from us, leaving the side which is towards us unillumined. In superior conjunction, on the other hand, the light falls full upon the side of the planet facing us; but the disc is then so small-looking, and our view besides is so dazzled by the proximity of the sun, that observations are of little avail. In the elongations, however, the sunlight comes from the side, and so we see one half of the planet lit up; the right half at eastern elongation, and the left half at western elongation. Piecing together the results given us at these more favourable views, we are enabled, bit by bit, to gather some small knowledge concerning the surface of an inferior planet.

From these considerations it will be seen at once that the inferior planets show various phases comparable to the waxing and waning of our moon in its monthly round. Superior conjunction is, in fact, similar to full moon, and inferior conjunction to new moon; while the eastern and western elongations may be compared respectively to the moon's first and last quarters. It will be recollected how, when these phases were first seen by the early telescopic observers, the Copernican theory was felt to be immensely strengthened; for it had been pointed out that if this system were the correct one, the planets Venus and Mercury, were it possible to see them more distinctly, would of necessity present phases like these when viewed from the earth. It should here be noted that the telescope was not invented until nearly seventy years after the death of Copernicus.

The apparent swing of an inferior planet from side to side of the sun, at one time on the east side, then passing into and lost in the sun's rays to appear once more on the west side, is the explanation of what is meant when we speak of an evening or a morning star. An inferior planet is called an evening star when it is at its eastern elongation, that is to say, on the left-hand of the sun; for, being then on the eastern side, it will set after the sun sets, as both sink in their turn below the western horizon at the close of day. Similarly, when such a planet is at its western elongation, that is to say, to the right-hand of the sun, it will go in advance of him, and so will rise above the eastern horizon before the sun rises, receiving therefore the designation of morning star. In very early times, however, before any definite ideas had been come to with regard to the celestial motions, it was generally believed that the morning and evening stars were quite distinct bodies. Thus Venus, when a morning star, was known to the ancients under the name of

Phosphorus, or Lucifer; whereas they called it Hesperus when it was an evening star.

Since an inferior planet circulates between us and the sun, one would be inclined to expect that such a body, each time it passed on the side nearest to the earth, should be seen as a black spot against the bright solar disc. Now this would most certainly be the case were the orbit of an inferior planet in the same plane with the orbit of the earth. But we have already seen how the orbits in the solar system, whether those of planets or of satellites, are by no means in the one plane; and that it is for this very reason that the moon is able to pass time after time in the direction of the sun, at the epoch known as new moon, and yet not to eclipse him save after the lapse of several such passages. Transits, then, as the passages of an inferior planet across the sun's disc are called, take place, for the same reason, only after certain regular lapses of time; and, as regards the circumstances of their occurrence, are on a par with eclipses of the sun. The latter, however, happen much more frequently, because the moon passes in the neighbourhood of the sun, roughly speaking, once a month, whereas Venus comes to each inferior conjunction at intervals so long apart as a year and a half, and Mercury only about every four months. From this it will be further gathered that transits of Mercury take place much oftener than transits of Venus.

Until recent years Transits of Venus were phenomena of great importance to astronomers, for they furnished the best means then available of calculating the distance of the sun from the earth. This was arrived at through comparing the amount of apparent displacement in the planet's path across the solar disc, when the transit was observed from widely separated stations on the earth's surface. The last transit of Venus took place in 1882, and there will not be another until the year 2004.

Transits of Mercury, on the other hand, are not of much scientific importance. They are of no interest as a popular spectacle; for the dimensions of the planet are so small, that it can be seen only with the aid of a telescope when it is in the act of crossing the sun's disc. The last transit of Mercury took place on November 14, 1907, and there will be another on November 6, 1914.

The first person known to have observed a transit of an inferior planet was the celebrated French philosopher, Gassendi. This was the transit of Mercury which took place on the 7th of December 1631.

The first time a transit of Venus was ever seen, so far as is known, was on the 24th of November 1639. The observer was a certain Jeremiah Horrox, curate of Hoole, near Preston, in Lancashire. The transit in question commenced shortly before sunset, and his observations in consequence were limited to only about half-an-hour. Horrox happened to have a great friend, one William Crabtree, of Manchester, whom he had advised by letter to be on the look out for the phenomenon. The weather in Crabtree's neighbourhood was cloudy, with the result that he only got a view of the transit for about ten minutes before the sun set.

That this transit was observed at all is due entirely to the remarkable ability of Horrox. According to the calculations of the great Kepler, no transit could take place that year (1639), as the planet would just pass clear of the lower edge of the sun. Horrox, however, not being satisfied with this, worked the question out for himself, and came to the conclusion that the planet would actually traverse the lower portion of the sun's disc. The event, as we have seen, proved him to be quite in the right. Horrox is said to have been a veritable prodigy of astronomical skill; and had he lived longer would, no doubt, have become very famous. Unfortunately he died about two years after his celebrated transit, in his twenty-second year only, according to the accounts. His friend Crabtree, who was then also a young man, is said to have been killed at the battle of Naseby in 1645.

There is an interesting phenomenon in connection with transits which is known as the "Black Drop." When an inferior planet has just made its way on to the face of the sun, it is usually seen to remain for a short time as if attached to the sun's edge by what looks like a dark ligament (see Fig. 12, p. 153). This gives to the planet for the time being an elongated appearance, something like that of a pear; but when the ligament, which all the while keeps getting thinner and thinner, has at last broken, the black body of the planet is seen to stand out round against the solar disc.

This appearance may be roughly compared to the manner in which a drop of liquid (or, preferably, of some glutinous substance) tends for a while to adhere to an object from which it is falling.

When the planet is in turn making its way off the face of the sun, the ligament is again seen to form and to attach it to the sun's edge before its due time.

The phenomenon of the black drop, or ligament, is entirely an illusion, and, broadly speaking, of an optical origin. Something very similar will be noticed if one brings one's thumb and forefinger slowly together against a very bright background.

This peculiar phenomenon has proved one of the greatest drawbacks to the proper observation of transits, for it is quite impossible to note the exact instant of the planet's entrance upon and departure from the solar disc in conditions such as these.

The black drop seems to bear a family resemblance, so to speak, to the phenomenon of Baily's beads. In the latter instance the lunar peaks, as they approach the sun's edge, appear to lengthen out in a similar manner and bridge the intervening space before their time, thus giving prominence to an effect which otherwise should scarcely be noticeable.

The last transit of Mercury, which, as has been already stated, took place on November 14, 1907, was not successfully observed by astronomers in England, on account of the cloudiness of the weather. In France, however, Professor Moye, of Montpellier, saw it under good conditions, and mentions that the black drop remained very conspicuous for fully a minute. The transit was also observed in the United States, the reports from which speak of the black drop as very "troublesome."

Before leaving the subject of transits it should be mentioned that it was in the capacity of commander of an expedition to Otaheite, in the Pacific, to observe the transit of Venus of June 3, 1769, that Captain Cook embarked upon the first of his celebrated voyages.

In studying the surfaces of Venus and Mercury with the telescope, observers are, needless to say, very much hindered by the proximity of the sun. Venus, when at the greatest elongations, certainly draws some distance out of the glare; but her surface is, even then, so dazzlingly bright, that the markings upon it are difficult to see. Mercury, on the other hand, is much duller in contrast, but the disc it shows in the telescope is exceedingly small; and, in addition, when that planet is left above the horizon for a short time after sunset, as necessarily happens after certain intervals, the mists near the earth's surface render observation of it very difficult.

Until about twenty-five years ago, it was generally believed that both these planets rotated on their axes in about twenty-four hours, a notion, no doubt, originally founded upon an unconscious desire to bring them into some conformity with our earth. But Schiaparelli, observing in Italy, and Percival Lowell, in the clear skies of Arizona and Mexico, have lately come to the conclusion that both planets rotate upon their axes in the same time as they revolve in their orbits,[12] the result being that they turn one face ever towards the sun in the same manner that the moon turns one face ever towards the earth--a curious state of things, which will be dealt with more fully when we come to treat of our satellite.

The marked difference in the brightness between the two planets has already been alluded to. The surface of Venus is, indeed, about five times as bright as that of Mercury. The actual brightness of Mercury is about equivalent to that of our moon, and astronomers are, therefore, inclined to think that it may resemble her in having a very rugged surface and practically no atmosphere. This probable lack of atmosphere is further corroborated by two circumstances. One of these is that when Mercury is just about to transit the face of the sun, no ring of diffused light is seen to encircle its disc as would be the case if it possessed an atmosphere. Such a lack of atmosphere is, indeed, only to be expected from what is known as the Kinetic Theory of Gases. According to this theory, which is based upon the behaviour of various kinds of gas, it is found that these elements tend to escape into space from the surface of bodies whose force of gravitation is weak. Hydrogen gas, for example, tends to fly away from our earth, as any one may see for himself when a balloon rises into the air. The gravitation of the earth seems, however, powerful enough to hold down other gases, as, for instance, those of which the air is chiefly composed, namely, oxygen and nitrogen. In due accordance with the Kinetic theory, we find the moon and Mercury, which are much about the same size, destitute of atmospheres. Mars, too, whose diameter is only about double that of the moon, has very little atmosphere. We find, on the other hand, that Venus, which is about the same size as our earth, clearly possesses an atmosphere, as just before the planet is in transit across the sun, the outline of its dark body is seen to be surrounded by a bright ring of light.

The results of telescopic observation show that more markings are visible on Mercury than on Venus. The intense brilliancy of Venus is, indeed, about the same as that of our white clouds when the sun is shining directly upon them. It has, therefore, been supposed that the planet is thickly enveloped in cloud, and that we do not ever see any part of its surface, except perchance the summit of some lofty mountain projecting through the fleecy mass.

With regard to the great brilliancy of Venus, it may be mentioned that she has frequently been seen in England, with the naked eye in full sunshine, when at the time of her greatest brightness. The writer has seen her thus at noonday. Needless to say, the sky at the moment was intensely blue and clear.

The orbit of Mercury is very oval, and much more so than that of any other planet. The consequence is that, when Mercury is nearest to the sun, the heat which it receives is twice as great as when it is farthest away. The orbit of Venus, on the other hand, is in marked contrast with that of Mercury, and is, besides, more nearly of a circular shape than that of any of the other planets. Venus, therefore, always keeps about the same distance from the sun, and so the heat which she receives during the course of her year can only be subject to very slight variations.

[11] In employing the terms Inferior and Superior the writer bows to astronomical custom, though he cannot help feeling that, in the circumstances, Interior and Exterior would be much more appropriate.

[12] This question is, however, uncertain, for some very recent spectroscopic observations of Venus seem to show a rotation period of about twenty-four hours.

CHAPTER XV

THE EARTH

We have already seen (in Chapter I.) how, in very early times, men naturally enough considered the earth to be a flat plane extending to a very great distance in every direction; but that, as years went on, certain of the Greek philosophers suspected it to be a sphere. One or two of the latter are, indeed, said to have further believed in its rotation about an axis, and even in its revolution around the sun; but, as the ideas in question were founded upon fancy, rather than upon any direct evidence, they did not generally attract attention. The small effect, therefore, which these theories had upon astronomy, may well be gathered from the fact that in the Ptolemaic system the earth was considered as fixed and at the centre of things; and this belief, as we have seen, continued unaltered down to the days of Copernicus. It was, indeed, quite impossible to be certain of the real shape of the earth or the reality of its motions until knowledge became more extended and scientific instruments much greater in precision.

We will now consider in detail a few of the more obvious arguments which can be put forward to show that our earth is a sphere.

If, for instance, the earth were a plane surface, a ship sailing away from us over the sea would appear to grow smaller and smaller as it receded into the distance, becoming eventually a tiny speck, and fading gradually from our view. This, however, is not at all what actually takes place. As we watch a vessel receding, its hull appears bit by bit to slip gently down over the horizon, leaving the masts alone visible. Then, in their turn, the masts are seen to slip down in the same manner, until eventually every trace of the vessel is gone. On the other hand, when a ship comes into view, the masts are the first portions to appear. They gradually rise up from below the horizon, and the hull follows in its turn, until the whole vessel is visible. Again, when one is upon a ship at sea, a set of masts will often be seen sticking up alone above the horizon, and these may shorten and gradually disappear from view without the body of the ship to which they belong becoming visible at all. Since one knows from experience that there is no edge at the horizon over which a vessel can drop down, the appearance which we have been describing can only be explained by supposing that the surface of the earth is always curving gradually in every direction.

The distance at which what is known as the horizon lies away from us depends entirely upon the height above the earth's surface where we happen at the moment to be. A ship which has appeared to sink below the horizon for a person standing on the beach, will be found to come back again into view if he at once ascends a high hill. Experiment shows that the horizon line lies at about three miles away for a person standing at the water's edge. The curving of the earth's surface is found, indeed, to be at the rate of eight inches in every mile. Now it can be ascertained, by calculation, that a body curving at this rate in every direction must be a globe about 8000 miles in diameter.

Again, the fact that, if not stopped by such insuperable obstacles as the polar ice and snow, those who travel continually in any one direction upon the earth's surface always find themselves back again at the regions from which they originally set out, is additional ground for concluding that the earth is a globe.

We can find still further evidence. For instance, in an eclipse of the moon the earth's shadow, when seen creeping across the moon's face, is noted to be always circular in shape. One cannot imagine how such a thing could take place unless the earth were a sphere.

Also, it is found from observation that the sun, the planets, and the satellites are, all of them, round. This roundness cannot be the roundness of a flat plate, for instance, for then the objects in question would sometimes present their thin sides to our view. It happens, also, that upon the discs which these bodies show, we see certain markings shifting along continually in one direction, to disappear at one side and to reappear again at the other. Such bodies must, indeed, be spheres in rotation.

The crescent and other phases, shown by the moon and the inferior planets, should further impress the truth of the matter upon us, as such appearances can only be caused by the sunlight falling from various directions upon the surfaces of spherical bodies.

Another proof, perhaps indeed the weightiest of all, is the continuous manner in which the stars overhead give place to others as one travels about the surface of the earth. When in northern regions the Pole Star and its neighbours--the stars composing the Plough, for instance--are over our heads. As one journeys south these gradually sink towards the northern horizon, while other stars take their place, and yet others are uncovered to view from the south. The regularity with which these changes occur shows that every point on the earth's surface faces a different direction of the sky, and such an arrangement would only be possible if the earth were a sphere. The celebrated Greek philosopher, Aristotle, is known to have believed in the globular shape of the earth, and it was by this very argument that he had convinced himself that it was so.

The idea of the sphericity of the earth does not appear, however, to have been generally accepted until the voyages of the great navigators showed that it could be sailed round.

The next point we have to consider is the rotation of the earth about its axis. From the earliest times men noticed that the sky and everything in it appeared to revolve around the earth in one fixed direction, namely, towards what is called the West, and that it made one complete revolution in the period of time which we know as twenty-four hours. The stars were seen to come up, one after another, from below the eastern horizon, to mount the sky, and then to sink in turn below the western horizon. The sun was seen to perform exactly the same journey, and the moon, too, whenever she was visible. One or two of the ancient Greek philosophers perceived that this might be explained, either by a movement of the entire heavens around the earth, or by a turning motion on the part of the earth itself. Of these diverse explanations, that which supposed an actual movement of the heavens appealed to them the most, for they could hardly conceive that the earth should continually rotate and men not be aware of its movement. The question may be compared to what we experience when borne along in a railway train. We see the telegraph posts and the trees and buildings near the line fly past us one after another in the contrary direction. Either these must be moving, or we must be moving; and as we happen to know that it is, indeed, we who are moving, there can be no question therefore about the matter. But it would not be at all so easy to be sure of this movement were one unable to see the objects close at hand displacing themselves. For instance, if one is shut up in a railway carriage at night with the blinds down, there is really nothing to show that one is moving, except the jolting of the train. And even then it is hard to be sure in which direction one is actually travelling.

The way we are situated upon the earth is therefore as follows. There are no other bodies sufficiently near to be seen flying past us in turn; our earth spins without a jolt; we and all things around us, including the atmosphere itself, are borne along together with precisely the same impetus, just as all the objects scattered about a railway carriage share in the forward movement of the train. Such being the case, what wonder that we are unconscious of the earth's rotation, of which we should know nothing at all, were it not for that slow displacement of the distant objects in the heavens, as we are borne past them in turn.

If the night sky be watched, it will be soon found that its apparent turning movement seems to take place around a certain point, which appears as if fixed. This point is known as the north pole of the heavens; and a rather bright star, which is situated very close to this hub of movement, is in consequence called the Pole Star. For the dwellers in southern latitudes there is also a point in their sky which appears to remain similarly fixed, and this is known as the south pole of the heavens. Since, however, the heavens do not turn round at all, but the earth does, it will easily be seen that these apparently stationary regions in the sky are really the points towards which the axis of the earth is directed. The positions on the earth's surface itself, known as the North and South Poles, are merely the places where the earth's axis, if there were actually such a thing, would be expected to jut out. The north pole of the earth will thus be situated exactly beneath the north pole of the heavens, and the south pole of the earth exactly beneath the south pole of the heavens.

We have seen that the earth rotates upon its imaginary axis once in about every twenty-four hours. This means that everything upon the surface of the earth is carried round once during that time. The measurement around the earth's equator is about 24,000 miles; and, therefore, an object situated at the equator must be carried round through a distance of about 24,000 miles in each twenty-four hours. Everything at the equator is thus moving along at the rapid rate of about 1000 miles an hour, or between sixteen and seventeen times as fast as an express train. If, however, one were to take measurements around the earth parallel to the equator, one would find these measurements becoming less and less, according as the poles were approached. It is plain, therefore, that the speed with which any point moves, in consequence of the earth's rotation, will be greatest at the equator, and less and less in the direction of the poles; while at the poles themselves there will be practically no movement, and objects there situated will merely turn round.

The considerations above set forth, with regard to the different speeds at which different portions of a rotating globe will necessarily be moving, is the foundation of an interesting experiment, which gives us further evidence of the rotation of our earth. The measurement around the earth at any distance below the surface, say, for instance, at the depth of a mile, will clearly be less than a similar measurement at the surface itself. The speed of a point at the bottom of a mine, which results from the actual rotation of the earth, must therefore be less than the speed of a point at the surface overhead. This can be definitely proved by dropping a heavy object down a mine shaft. The object, which starts with the greater speed of the surface, will, when it reaches the bottom of the mine, be found, as might be indeed expected, to be a little ahead (i.e. to the east) of the point which originally lay exactly underneath it. The distance by which the object gains upon this point is, however, very small. In our latitudes it amounts to about an inch in a fall of 500 feet.

The great speed at which, as we have seen, the equatorial regions of the earth are moving, should result in giving to the matter there situated a certain tendency to fly outwards. Sir Isaac Newton was the first to appreciate this point, and he concluded from it that the earth must be bulged a little all round the equator. This is, indeed, found to be the case, the diameter at the equator being nearly twenty-seven miles greater than it is from pole to pole. The reader will, no doubt, be here reminded of the familiar comparison in geographies between the shape of the earth and that of an orange.

In this connection it is interesting to consider that, were the earth to rotate seventeen times as fast as it does (i.e. in one hour twenty-five minutes, instead of twenty-four hours), bodies at the equator would have such a strong tendency to fly outwards that the force of terrestrial gravity acting upon them would just be counterpoised, and they would virtually have no weight. And, further, were the earth to rotate a little faster still, objects lying loose upon its surface would be shot off into space.

The earth is, therefore, what is technically known as an oblate spheroid; that is, a body of spherical shape flattened at the poles. It follows of course from this, that objects at the polar regions are slightly nearer to the earth's centre than objects at the equatorial regions. We have already seen that gravitation acts from the central parts of a body, and that its force is greater the nearer are those central parts. The result of this upon our earth will plainly be that objects in the polar regions will be pulled with a slightly stronger pull, and will therefore weigh a trifle more than objects in the equatorial regions. This is, indeed, found by actual experiment to be the case. As an example of the difference in question, Professor Young, in his Manual of Astronomy, points out that a man who weighs 190 pounds at the equator would weigh 191 at the pole. In such an experiment the weighing would, however, have to be made with a spring balance, and not with scales; for, in the latter case, the "weights" used would alter in their weight in exactly the same degree as the objects to be weighed.

It used to be thought that the earth was composed of a relatively thin crust, with a molten interior. Scientific men now believe, on the other hand, that such a condition cannot after all prevail, and that the earth must be more or less solid all through, except perhaps in certain isolated places where collections of molten matter may exist.

The atmosphere, or air which we breathe, is in the form of a layer of limited depth which closely envelops the earth. Actually, it is a mixture of several gases, the most important being nitrogen and oxygen, which between them practically make up the air, for the proportion of the other gases, the chief of which is carbonic acid gas, is exceedingly small.

It is hard to picture our earth, as we know it, without this atmosphere. Deprived of it, men at once would die; but even if they could be made to go on living without it by any miraculous means, they would be like unto deaf beings, for they would never hear any sound. What we call sounds are merely vibrations set up in the air, which travel along and strike upon the drum of the ear.

The atmosphere is densest near the surface of the earth, and becomes less and less dense away from it, as a result of diminishing pressure of air from above. The greater portion of it is accumulated within four or five miles of the earth's surface.

It is impossible to determine exactly at what distance from the earth's surface the air ceases altogether, for it grows continually more and more rarefied. There are, however, two distinct methods of ascertaining the distance beyond which it can be said practically not to exist. One of these methods we get from twilight. Twilight is, in fact, merely light reflected to us from those upper regions of the air, which still continue to be illuminated by the sun after it has disappeared from our view below the horizon. The time during which twilight lasts, shows us that the atmosphere must be at least fifty miles high.

But the most satisfactory method of ascertaining the height to which the atmosphere extends is from the observation of meteors. It is found that these bodies become ignited, by the friction of passing into the atmosphere, at a height of about 100 miles above the surface of the earth. We thus gather that the atmosphere has a certain degree of density even at this height. It may, indeed, extend as far as about 150 miles.

The layer of atmosphere surrounding our earth acts somewhat in the manner of the glass covering of a greenhouse, bottling in the sun's rays, and thus storing up their warmth for our benefit. Were this not so, the heat which we get from the sun would, after falling upon the earth, be quickly radiated again into space.

It is owing to the unsteadiness of the air that stars are seen to twinkle. A night when this takes place, though it may please the average person, is worse than useless to the astronomer, for the unsteadiness is greatly magnified in the telescope. This twinkling is, no doubt, in a great measure responsible for the conventional "points" with which Art has elected to embellish stars, and which, of course, have no existence in fact.

The phenomena of Refraction,[13] namely, that bending which rays of light undergo, when passing slant-wise from a rare into a dense transparent medium, are very marked with regard to the atmosphere. The denser the medium into which such rays pass, the greater is this bending found to be. Since the layer of air around us becomes denser and denser towards the surface of the earth, it will readily be granted that the rays of light reaching our eyes from a celestial object, will suffer the greater bending the lower the object happens to be in the sky. Celestial objects, unless situated directly overhead, are thus not seen in their true places, and when nearest to the horizon are most out of place. The bending alluded to is upwards. Thus the sun and the moon, for instance, when we see them resting upon the horizon, are actually entirely beneath it.

When the sun, too, is sinking towards the horizon, the lower edge of its disc will, for the above reason, look somewhat more raised than the upper. The result is a certain appearance of flattening; which may plainly be seen by any one who watches the orb at setting.

In observations to determine the exact positions of celestial objects correction has to be made for the effects of refraction, according to the apparent elevation of these objects in the sky. Such effects are least when the objects in question are directly overhead, for then the rays of light, coming from them to the eye, enter the atmosphere perpendicularly, and not at any slant.

A very curious effect, due to refraction, has occasionally been observed during a total eclipse of the moon. To produce an eclipse of this kind, the earth must, of course, lie directly between the sun and the moon. Therefore, when we see the shadow creeping over the moon's surface, the sun should actually be well below the horizon. But when a lunar eclipse happens to come on just about sunset, the sun, although really sunk below the horizon, appears still above it through refraction, and the eclipsed moon, situated, of course, exactly opposite to it in the sky, is also lifted up above the horizon by the same cause. Pliny, writing in the first century of the Christian era, describes an eclipse of this kind, and refers to it as a "prodigy." The phenomenon is known as a "horizontal eclipse." It was, no doubt, partly owing to it that the ancients took so long to decide that an eclipse of the moon was really caused by the shadow cast by the earth. Plutarch, indeed, remarks that it was easy enough to understand that a solar eclipse was caused by the interposition of the moon, but that one could not imagine by the interposition of what body the moon itself could be eclipsed.

In that apparent movement of the heavens about the earth, which men now know to be caused by the mere rotation of the earth itself, a slight change is observed to be continually taking place. The stars, indeed, are always found to be gradually drawing westward, i.e. towards the sun, and losing themselves one after the other in the blaze of his light, only to reappear, however, on the other side of him after a certain lapse of time. This is equivalent to saying that the sun itself seems always creeping slowly eastward in the heaven. The rate at which this appears to take place is such that the sun finds itself back again to its original position, with regard to the starry background, at the end of a year's time. In other words, the sun seems to make a complete tour of the heavens in the course of a year. Here, however, we have another illusion, just as the daily movement of the sky around the earth was an illusion. The truth indeed is, that this apparent movement of the sun eastward among the stars during a year, arises merely from a continuous displacement of his position caused by an actual motion of the earth itself around him in that very time. In a word, it is the earth which really moves around the sun, and not the sun around the earth.

The stress laid upon this fundamental point by Copernicus, marks the separation of the modern from the ancient view. Not that Copernicus, indeed, had obtained any real proof that the earth is merely a planet revolving around the sun; but it seemed to his profound intellect that a movement of this kind on the part of our globe was the more likely explanation of the celestial riddle. The idea was not new; for, as we have already seen, certain of the ancient Greeks (Aristarchus of Samos, for example) had held such a view; but their notions on the subject were very fanciful, and unsupported by any good argument.

What Copernicus, however, really seems to have done was to insist upon the idea that the sun occupied the centre, as being more consonant with common sense. No doubt, he was led to take up this position by the fact that the sun appeared entirely of a different character from the other members of the system. The one body in the scheme, which performed the important function of dispenser of light and heat, would indeed be more likely to occupy a position apart from the rest; and what position more appropriate for its purposes than the centre!

But here Copernicus only partially solved the difficult question. He unfortunately still clung to an ancient belief, which as yet remained unquestioned; i.e. the great virtue, one might almost say, the divineness, of circular motion. The ancients had been hag-ridden, so to speak, by the circle; and it appeared to them that such a perfectly formed curve was alone fitted for the celestial motions. Ptolemy employed it throughout his system. According to him the "planets" (which included, under the ancient view, both the sun and the moon), moved around the earth in circles; but, as their changing positions in the sky could not be altogether accounted for in this way, it was further supposed that they performed additional circular movements, around peculiarly placed centres, during the course of their orbital revolutions. Thus the Ptolemaic system grew to be extremely complicated; for astronomers did not hesitate to add new circular movements whenever the celestial positions calculated for the planets were found not to tally with the positions observed. In this manner, indeed, they succeeded in doctoring the theory, so that it fairly satisfied the observations made with the rough instruments of pre-telescopic times.

Although Copernicus performed the immense service to astronomy of boldly directing general attention to the central position of the sun, he unfortunately took over for the new scheme the circular machinery of the Ptolemaic system. It therefore remained for the famous Kepler, who lived about a century after him, to find the complete solution. Just as Copernicus, for instance, had broken free from tradition with regard to the place of the sun; so did Kepler, in turn, break free from the spell of circular motion, and thus set the coping-stone to the new astronomical edifice. This astronomer showed, in fact, that if the paths of the planets around the sun, and of the moon around the earth, were not circles, but ellipses, the movements of these bodies about the sky could be correctly accounted for. The extreme simplicity of such an arrangement was far more acceptable than the bewildering intricacy of movement required by the Ptolemaic theory. The Copernican system, as amended by Kepler, therefore carried the day; and was further strengthened, as we have already seen, by the telescopic observations of Galileo and the researches of Newton into the effects of gravitation.

And here a word on the circle, now fallen from its high estate. The ancients were in error in supposing that it stood entirely apart--the curve of curves. As a matter of fact it is merely a special kind of ellipse. To put it paradoxically, it is an ellipse which has no ellipticity, an oval without any ovalness!

Notwithstanding all this, astronomy had to wait yet a long time for a definite proof of the revolution of the earth around the sun. The leading argument advanced by Aristotle, against the reality of any movement of the earth, still held good up to about seventy years ago. That philosopher had pointed out that the earth could not move about in space to any great extent, or the stars would be found to alter their apparent places in the sky, a thing which had never been observed to happen. Centuries ran on, and instruments became more and more perfect, yet no displacements of stars were noted. In accepting the Copernican theory men were therefore obliged to suppose these objects as immeasurably distant. At length, however, between the years 1835 and 1840, it was discovered by the Prussian astronomer, Bessel, that a star known as 61 Cygni--that is to say, the star marked in celestial atlases as No. 61 in the constellation of the Swan--appeared, during the course of a year, to perform a tiny circle in the heavens, such as would result from a movement on our own part around the sun. Since then about forty-three stars have been found to show minute displacements of a similar kind, which cannot be accounted for upon any other supposition than that of a continuous revolution of the earth around the sun. The triumph of the Copernican system is now at last supreme.

If the axis of the earth stood "straight up," so to speak, while the earth revolved in its orbit, the sun would plainly keep always on a level with the equator. This is equivalent to stating that, in such circumstances, a person at the equator would see it rise each morning exactly in the east, pass through the zenith, that is, the point directly overhead of him, at midday, and set in the evening due in the west. As this would go on unchangingly at the equator every day throughout the year, it should be clear that, at any particular place upon the earth, the sun would in these conditions always be seen to move in an unvarying manner across the sky at a certain altitude depending upon the latitude of the place. Thus the more north one went upon the earth's surface, the more southerly in the sky would the sun's path lie; while at the north pole itself, the sun would always run round and round the horizon. Similarly, the more south one went from the equator the more northerly would the path of the sun lie, while at the south pole it would be seen to skirt the horizon in the same manner as at the north pole. The result of such an arrangement would be, that each place upon the earth would always have one unvarying climate; in which case there would not exist any of those beneficial changes of season to which we owe so much.

The changes of season, which we fortunately experience, are due, however, to the fact that the sun does not appear to move across the sky each day at one unvarying altitude, but is continually altering the position of its path; so that at one period of the year it passes across the sky low down, and remains above the horizon for a short time only, while at another it moves high up across the heavens, and is above the horizon for a much longer time. Actually, the sun seems little by little to creep up the sky during one half of the year, namely, from mid-winter to mid-summer, and then, just as gradually, to slip down it again during the other half, namely, from mid-summer to mid-winter. It will therefore be clear that every region of the earth is much more thoroughly warmed during one portion of the year than during another, i.e. when the sun's path is high in the heavens than when it is low down.

Once more we find appearances exactly the contrary from the truth. The earth is in this case the real cause of the deception, just as it was in the other cases. The sun does not actually creep slowly up the sky, and then slowly dip down it again, but, owing to the earth's axis being set aslant, different regions of the earth's surface are presented to the sun at different times. Thus, in one portion of its orbit, the northerly regions of the earth are presented to the sun, and in the other portion the southerly. It follows of course from this, that when it is summer in the northern hemisphere it is winter in the southern, and vice vers?(see Fig. 13, p. 176).

The fact that, in consequence of this slant of the earth's axis, the sun is for part of the year on the north side of the equator and part of the year on the south side, leads to a very peculiar result. The path of the moon around the earth is nearly on the same plane with the earth's path around the sun. The moon, therefore, always keeps to the same regions of the sky as the sun. The slant of the earth's axis thus regularly displaces the position of both the sun and the moon to the north and south sides of the equator respectively in the manner we have been describing. Were the earth, however, a perfect sphere, such change of position would not produce any effect. We have shown, however, that the earth is not a perfect sphere, but that it is bulged out all round the equator. The result is that this bulged-out portion swings slowly under the pulls of solar and lunar gravitation, in response to the displacements of the sun and moon to the north and to the south of it. This slow swing of the equatorial regions results, of course, in a certain slow change of the direction of the earth's axis, so that the north pole does not go on pointing continually to the same region of the sky. The change in the direction of the axis is, however, so extremely slight, that it shows up only after the lapse of ages. The north pole of the heavens, that is, the region of the sky towards which the north pole of the earth's axis points, displaces therefore extremely slowly, tracing out a wide circle, and arriving back again to the same position in the sky only after a period of about 25,000 years. At present the north pole of the heavens is quite close to a bright star in the tail of the constellation of the Little Bear, which is consequently known as the Pole Star; but in early Greek times it was at least ten times as far away from this star as it is now. After some 12,000 years the pole will point to the constellation of Lyra, and Vega, the most brilliant star in that constellation, will then be considered as the pole star. This slow twisting of the earth's axis is technically known as Precession, or the Precession of the Equinoxes (see Plate XIX., p. 292).

The slow displacement of the celestial pole appears to have attracted the attention of men in very early times, but it was not until the second century B.C. that precession was established as a fact by the celebrated Greek astronomer, Hipparchus. For the ancients this strange cyclical movement had a mystic significance; and they looked towards the end of the period as the end, so to speak, of a "dispensation," after which the life of the universe would begin anew.

We have seen that the orbit of the earth is an ellipse, and that the sun is situated at what is called the focus, a point not in the middle of the ellipse, but rather towards one of its ends. Therefore, during the course of the year the distance of the earth from the sun varies. The sun, in consequence of this, is about 3,000,000 miles nearer to us in our northern winter than it is in our northern summer, a statement which sounds somewhat paradoxical. This variation in distance, large as it appears in figures, can, however, not be productive of much alteration in the amount of solar heat which we receive, for during the first week in January, when the distance is least, the sun only looks about one-eighteenth broader than at the commencement of July, when the distance is greatest. The great disparity in temperature between winter and summer depends, as we have seen, upon causes of quite another kind, and varies between such wide limits that the effects of this slight alteration in the distance of the sun from the earth may be neglected for practical purposes.

The Tides are caused by the gravitational pull of the sun and moon upon the water of the earth's surface. Of the two, the moon, being so much the nearer, exerts the stronger pull, and therefore may be regarded as the chief cause of the tides. This pull always draws that portion of the water, which happens to be right underneath the moon at the time, into a heap; and there is also a second heaping of water at the same moment at the contrary side of the earth, the reasons for which can be shown mathematically, but cannot be conveniently dealt with here.

As the earth rotates on its axis each portion of its surface passes beneath the moon, and is swelled up by this pull; the watery portions being, however, the only ones to yield visibly. A similar swelling up, as we have seen, takes place at the point exactly away from the moon. Thus each portion of our globe is borne by the rotation through two "tide-areas" every day, and this is the reason why there are two tides during every twenty-four hours.

The crest of the watery swelling is known as high tide. The journey of the moon around the earth takes about a month, and this brings her past each place in turn by about fifty minutes later each day, which is the reason why high tide is usually about twenty-five minutes later each time.

The moon is, however, not the sole cause of the tides, but the sun, as we have said, has a part in the matter also. When it is new moon the gravitational attractions of both sun and moon are clearly acting together from precisely the same direction, and, therefore, the tide will be pulled up higher than at other times. At full moon, too, the same thing happens; for, although the bodies are now acting from opposite directions, they do not neutralise each other's pulls as one might imagine, since the sun, in the same manner as the moon, produces a tide both under it and also at the opposite side of the earth. Thus both these tides are actually increased in height. The exceptionally high tides which we experience at new and full moons are known as Spring Tides, in contradistinction to the minimum high tides, which are known as Neap Tides.

The ancients appear to have had some idea of the cause of the tides. It is said that as early as 1000 B.C. the Chinese noticed that the moon exerted an influence upon the waters of the sea. The Greeks and Romans, too, had noticed the same thing; and Caesar tells us that when he was embarking his troops for Britain the tide was high because the moon was full. Pliny went even further than this, in recognising a similar connection between the waters and the sun.

From casual observation one is inclined to suppose that the high tide always rises many feet. But that this is not the case is evidenced by the fact that the tides in the midst of the great oceans are only from three to four feet high. However, in the seas and straits around our Isles, for instance, the tides rise very many feet indeed, but this is merely owing to the extra heaping up which the large volumes of water undergo in forcing their passage through narrow channels.

As the earth, in rotating, is continually passing through these tide-areas, one might expect that the friction thus set up would tend to slow down the rotation itself. Such a slowing down, or "tidal drag," as it is called, is indeed continually going on; but the effects produced are so exceedingly minute that it will take many millions of years to make the rotation appreciably slower, and so to lengthen the day.

Recently it has been proved that the axis of the earth is subject to a very small displacement, or rather, "wobbling," in the course of a period of somewhat over a year. As a consequence of this, the pole shifts its place through a circle of, roughly, a few yards in width during the time in question. This movement is, perhaps, the combined result of two causes. One of these is the change of place during the year of large masses of material upon our earth; such as occurs, for instance, when ice and snow melt, or when atmospheric and ocean currents transport from place to place great bodies of air and water. The other cause is supposed to be the fact that the earth is not absolutely rigid, and so yields to certain strains upon it. In the course of investigation of this latter point the interesting conclusion has been reached by the famous American astronomer, Professor Simon Newcomb, that our globe as a whole is a little more rigid than steel.

We will bring this chapter to a close by alluding briefly to two strange appearances which are sometimes seen in our night skies. These are known respectively as the Zodiacal Light and the Gegenschein.

The Zodiacal Light is a faint cone-shaped illumination which is seen to extend upwards from the western horizon after evening twilight has ended, and from the eastern horizon before morning twilight has begun. It appears to rise into the sky from about the position where the sun would be at that time. The proper season of the year for observing it during the evening is in the spring, while in autumn it is best seen in the early morning. In our latitudes its light is not strong enough to render it visible when the moon is full, but in the tropics it is reported to be very bright, and easily seen in full moonlight. One theory regards it as the reflection of light from swarms of meteors revolving round the sun; another supposes it to be a very rarefied extension of the corona.

The Gegenschein (German for "counter-glow") is a faint oval patch of light, seen in the sky exactly opposite to the place of the sun. It is usually treated of in connection with the zodiacal light, and one theory regards it similarly as of meteoric origin. Another theory, however--that of Mr. Evershed--considers it a sort of tail to the earth (like a comet's tail) composed of hydrogen and helium--the two lightest gases we know--driven off from our planet in the direction contrary to the sun.

[13] Every one knows the simple experiment in which a coin lying at the bottom of an empty basin, and hidden from the eye by its side, becomes visible when a certain quantity of water has been poured in. This is an example of refraction. The rays of light coming from the coin ought not to reach the eye, on account of the basin's side being in the way; yet by the action of the water they are refracted, or bent over its edge, in such a manner that they do.

CHAPTER XVI

THE MOON

What we call the moon's "phases" are merely the various ways in which we see the sun shining upon her surface during the course of her monthly revolutions around the earth (see Fig. 14, p. 184). When she passes in the neighbourhood of the sun all his light falls upon that side which is turned away from us, and so the side which is turned towards us is unillumined, and therefore invisible. When in this position the moon is spoken of as new.

As she continues her motion around the earth, she draws gradually to the east of the sun's place in the sky. The sunlight then comes somewhat from the side; and so we see a small portion of the right side of the lunar disc illuminated. This is the phase known as the crescent moon.

As she moves on in her orbit more and more of her illuminated surface is brought into view; and so the crescent of light becomes broader and broader, until we get what is called half-moon, or first quarter, when we see exactly one-half of her surface lit up by the sun's rays. As she draws still further round yet more of her illuminated surface is brought into view, until three-quarters of the disc appear lighted up. She is then said to be gibbous.

Eventually she moves round so that she faces the sun completely, and the whole of her disc appears illuminated. She is then spoken of as full. When in this position it is clear that she is on the contrary side of the earth to the sun, and therefore rises about the same time that he is setting. She is now, in fact, at her furthest from the sun.

After this, the motion of the moon in her orbit carries her on back again in the direction of the sun. She thus goes through her phases as before, only these of course are in the reverse order. The full phase is seen to give place to the gibbous, and this in turn to the half-moon and to the crescent; after which her motion carries her into the neighbourhood of the sun, and she is once more new, and lost to our sight in the solar glare. Following this she draws away to the east of the sun again, and the old order of phases repeat themselves as before.

The early Babylonians imagined that the moon had a bright and a dark side, and that her phases were caused by the bright side coming more and more into view during her movement around the sky. The Greeks, notably Aristotle, set to work to examine the question from a mathematical standpoint, and came to the conclusion that the crescent and other appearances were such as would necessarily result if the moon were a dark body of spherical shape illumined merely by the light of the sun.

Although the true explanation of the moon's phases has thus been known for centuries, it is unfortunately not unusual to see pictures--advertisement posters, for instance--in which stars appear within the horns of a crescent moon! Can it be that there are to-day educated persons who believe that the moon is a thing which grows to a certain size and then wastes away again; who, in fact, do not know that the entire body of the moon is there all the while?

When the moon shows a very thin crescent, we are able dimly to see her still dark portion standing out against the sky. This appearance is popularly known as the "old moon in the new moon's arms." The dark part of her surface must, indeed, be to some degree illumined, or we should not be able to see it at all. Whence then comes the light which illumines it, since it clearly cannot come from the sun? The riddle is easily solved, if we consider what kind of view of our earth an observer situated on this darkened part of the moon would at that moment get. He would, as a matter of fact, just then see nearly the whole disc of the earth brightly lit up by sunlight. The lunar landscape all around would, therefore, be bathed in what to him would be "earthlight," which of course takes the place there of what we call moonlight. If, then, we recollect how much greater in size the earth is than the moon, it should not surprise us that this earthlight will be many times brighter than moonlight. It is considered, indeed, to be some twenty times brighter. It is thus not at all astonishing that we can see the dark portion of the moon illumined merely by sunlight reflected upon it from our earth.

The ancients were greatly exercised in their minds to account for this "earthlight," or "earthshine," as it is also called. Posidonius (135-51 B.C.) tried to explain it by supposing that the moon was partially transparent, and that some sunlight consequently filtered through from the other side. It was not, however, until the fifteenth century that the correct solution was arrived at.

Perhaps the most remarkable thing which one notices about the moon is that she always turns the same side towards us, and so we never see her other side. One might be led from this to jump to the conclusion that she does not rotate upon an axis, as do the other bodies which we see; but, paradoxical as it may appear, the fact that she turns one face always towards the earth, is actually a proof that she does rotate upon an axis. The rotation, however, takes place with such slowness, that she turns round but once during the time in which she revolves around the earth (see Fig. 15). In order to understand the matter clearly, let the reader place an object in the centre of a room and walk around it once, keeping his face turned towards it the whole time, While he is doing this, it is evident that he will face every one of the four walls of the room in succession. Now in order to face each of the four walls of a room in succession one would be obliged to turn oneself entirely round. Therefore, during the act of walking round an object with his face turned directly towards it, a person at the same time turns his body once entirely round.

In the long, long past the moon must have turned round much faster than this. Her rate of rotation has no doubt been slowed down by the action of some force. It will be recollected how, in the course of the previous chapter, we found that the tides were tending, though exceedingly gradually, to slow down the rotation of the earth upon its axis. But, on account of the earth's much greater mass, the force of gravitation exercised by it upon the surface of the moon is, of course, much more powerful than that which the moon exercises upon the surface of the earth. The tendency to tidal action on the moon itself must, therefore, be much in excess of anything which we here experience. It is, in consequence, probable that such a tidal drag, extending over a very long period of time, has resulted in slowing down the moon's rotation to its present rate.

The fact that we never see but one side of the moon has given rise from time to time to fantastic speculations with regard to the other side. Some, indeed, have wished to imagine that our satellite is shaped like an egg, the more pointed end being directed away from us. We are here, of course, faced with a riddle, which is all the more tantalising from its appearing for ever insoluble to men, chained as they are to the earth. However, it seems going too far to suppose that any abnormal conditions necessarily exist at the other side of the moon. As a matter of fact, indeed, small portions of that side are brought into our view from time to time in consequence of slight irregularities in the moon's movement; and these portions differ in no way from those which we ordinarily see. On the whole, we obtain a view of about 60 per cent. of the entire lunar surface; that is to say, a good deal more than one-half.

The actual diameter of the moon is about 2163 miles, which is somewhat more than one-quarter the diameter of the earth. For a satellite, therefore, she seems very large compared with her primary, the earth; when we consider that Jupiter's greatest satellite, although nearly twice as broad as our moon, has a diameter only one twenty-fifth that of Jupiter. Furthermore, the moon moves around the earth comparatively slowly, making only about thirteen revolutions during the entire year. Seen from space, therefore, she would not give the impression of a circling body, as other satellites do. Her revolutions are, indeed, relatively so very slow that she would appear rather like a smaller planet accompanying the earth in its orbit. In view of all this, some astronomers are inclined to regard the earth and moon rather as a "double planet" than as a system of planet and satellite.

When the moon is full she attracts more attention perhaps than in any of her other phases. The moon, in order to be full, must needs be in that region of the heavens exactly opposite to the sun. The sun appears to go once entirely round the sky in the course of a year, and the moon performs the same journey in the space of about a month. The moon, when full, having got half-way round this journey, occupies, therefore, that region of the sky which the sun itself will occupy half a year later. Thus in winter the full moon will be found roughly to occupy the sun's summer position in the sky, and in summer the sun's winter position. It therefore follows that the full moon in winter time is high up in the heavens, while in summer time it is low down. We thus get the greatest amount of full moonlight when it is the most needed.

The great French astronomer, Laplace, being struck by the fact that the "lesser light" did not rule the night to anything like the same extent that the "greater light" ruled the day, set to work to examine the conditions under which it might have been made to do so. The result of his speculations showed that if the moon were removed to such a distance that she took a year instead of a month to revolve around the earth; and if she were started off in her orbit at full moon, she would always continue to remain full--a great advantage for us. Whewell, however, pointed out that in order to get the moon to move with the requisite degree of slowness, she would have to revolve so far from the earth that she would only look one-sixteenth as large as she does at present, which rather militates against the advantage Laplace had in mind! Finally, however, it was shown by M. Liouville, in 1845, that the position of a perennial full moon, such as Laplace dreamed of, would be unstable--that is to say, the body in question could not for long remain undisturbed in the situation suggested (see Fig. 16, p. 191).

There is a well-known phenomenon called harvest moon, concerning the nature of which there seems to be much popular confusion. An idea in fact appears to prevail among a good many people that the moon is a harvest moon when, at rising, it looks bigger and redder than usual. Such an appearance has, however, nothing at all to say to the matter; for the moon always looks larger when low down in the sky, and, furthermore, it usually looks red in the later months of the year, when there is more mist and fog about than there is in summer. What astronomers actually term the harvest moon is, indeed, something entirely different from this. About the month of September the slant at which the full moon comes up from below the horizon happens to be such that, during several evenings together, she rises almost at the same hour, instead of some fifty minutes later, as is usually the case. As the harvest is being gathered in about that time, it has come to be popularly considered that this is a provision of nature, according to which the sunlight is, during several evenings, replaced without delay by more or less full-moonlight, in order that harvesters may continue their work straight on into the night, and not be obliged to break off after sunset to wait until the moon rises. The same phenomenon is almost exactly repeated a month later, but by reason of the pursuits then carried on it is known as the "hunter's moon."

In this connection should be mentioned that curious phenomenon above alluded to, and which seems to attract universal notice, namely, that the moon looks much larger when near the horizon--at its rising, for instance, than when higher up in the sky. This seeming enlargement is, however, by no means confined to the moon. That the sun also looks much larger when low down in the sky than when high up, seems to strike even the most casual watcher of a sunset. The same kind of effect will, indeed, be noted if close attention be paid to the stars themselves. A constellation, for instance, appears more spread out when low down in the sky than when high up. This enlargement of celestial objects when in the neighbourhood of the horizon is, however, only apparent and not real. It must be entirely an illusion; for the most careful measurements of the discs of the sun and of the moon fail to show that the bodies are any larger when near the horizon than when high up in the sky. In fact, if there be any difference in measurements with regard to the moon, it will be found to be the other way round; for her disc, when carefully measured, is actually the slightest degree greater when high in the sky, than when low down. The reason for this is that, on account of the rotundity of the earth's surface, she is a trifle nearer the observer when overhead of him.

This apparent enlargement of celestial objects, when low down in the sky, is granted on all sides to be an illusion; but although the question has been discussed with animation time out of mind, none of the explanations proposed can be said to have received unreserved acceptance. The one which usually figures in text-books is that we unconsciously compare the sun and moon, when low down in the sky, with the terrestrial objects in the same field of view, and are therefore inclined to exaggerate the size of these orbs. Some persons, on the other hand, imagine the illusion to have its source in the structure of the human eye; while others, again, put it down to the atmosphere, maintaining that the celestial objects in question loom large in the thickened air near the horizon, in the same way that they do when viewed through fog or mist.

The writer[14] ventures, however, to think that the illusion has its origin in our notion of the shape of the celestial vault. One would be inclined, indeed, to suppose that this vault ought to appear to us as the half of a hollow sphere; but he maintains that it does not so appear, as a consequence of the manner in which the eyes of men are set quite close together in their heads. If one looks, for instance, high up in the sky, the horizon cannot come within the field of view, and so there is nothing to make one think that the expanse then gazed upon is other than quite flat--let us say like the ceiling of a room. But, as the eyes are lowered, a portion of the encircling horizon comes gradually into the field of view, and the region of the sky then gazed upon tends in consequence to assume a hollowed-out form. From this it would seem that our idea of the shape of the celestial vault is, that it is flattened down over our heads and hollowed out all around in the neighbourhood of the horizon (see Fig. 17, p. 195). Now, as a consequence of their very great distance, all the objects in the heavens necessarily appear to us to move as if they were placed on the background of the vault; the result being that the mind is obliged to conceive them as expanded or contracted, in its unconscious attempts to make them always fill their due proportion of space in the various parts of this abnormally shaped sky.

From such considerations the writer concludes that the apparent enlargement in question is merely the natural consequence of the idea we have of the shape of the celestial vault--an idea gradually built up in childhood, to become later on what is called "second nature." And in support of this contention, he would point to the fact that the enlargement is not by any means confined to the sun and moon, but is every whit as marked in the case of the constellations. To one who has not noticed this before, it is really quite a revelation to compare the appearance of one of the large constellations (Orion, for instance) when high up in the sky and when low down. The widening apart of the various stars composing the group, when in the latter position, is very noticeable indeed.

Further, if a person were to stand in the centre of a large dome, he would be exactly situated as if he were beneath the vaulted heaven, and one would consequently expect him to suffer the same illusion as to the shape of the dome. Objects fixed upon its background would therefore appear to him under the same conditions as objects in the sky, and the illusions as to their apparent enlargement should hold good here also.

Some years ago a Belgian astronomer, M. Stroobant, in an investigation of the matter at issue, chanced to make a series of experiments under the very conditions just detailed. To various portions of the inner surface of a large dome he attached pairs of electric lights; and on placing himself at the centre of the building, he noticed that, in every case, those pairs which were high up appeared closer together than those which were low down! He does not, however, seem to have sought for the cause in the vaulted expanse. On the contrary, he attributed the effect to something connected with our upright stature, to some physiological reason which regularly makes us estimate objects as larger when in front than when overhead.

In connection with this matter, it may be noted that it always appears extremely difficult to estimate with the eye the exact height above the horizon at which any object (say a star) happens to be. Even skilled observers find themselves in error in attempting to do so. This seems to bear out the writer's contention that the form under which the celestial vault really appears to us is a peculiar one, and tends to give rise to false judgments.

Before leaving this question, it should also be mentioned that nothing perhaps is more deceptive than the size which objects in the sky appear to present. The full moon looks so like a huge plate, that it astonishes one to find that a threepenny bit held at arm's length will a long way more than cover its disc.

The moon is just too far off to allow us to see the actual detail on her surface with the naked eye. When thus viewed she merely displays a patchy appearance,[15] and the imaginary forms which her darker markings suggest to the fancy are popularly expressed by the term "Man in the Moon." An examination of her surface with very moderate optical aid is, however, quite a revelation, and the view we then get is not easily comparable to what we see with the unaided eye.

Even with an ordinary opera-glass, an observer will be able to note a good deal of detail upon the lunar disc. If it be his first observation of the kind, he cannot fail to be struck by the fact to which we have just made allusion, namely, the great change which the moon appears to undergo when viewed with magnifying power. "Cain and his Dog," the "Man in the Moon gathering sticks," or whatever indeed his fancy was wont to conjure up from the lights and shades upon the shining surface, have now completely disappeared; and he sees instead a silvery globe marked here and there with extensive dark areas, and pitted all over with crater-like formations (see Plate VIII., p. 196). The dark areas retain even to the present day their ancient name of "seas," for Galileo and the early telescopic observers believed them to be such, and they are still catalogued under the mystic appellations given to them in the long ago; as, for instance, "Sea of Showers," "Bay of Rainbows," "Lake of Dreams."[16] The improved telescopes of later times showed, however, that they were not really seas (there is no water on the moon), but merely areas of darker material.

The crater-like formations above alluded to are the "lunar mountains." A person examining the moon for the first time with telescopic aid, will perhaps not at once grasp the fact that his view of lunar mountains must needs be what is called a "bird's-eye" one, namely, a view from above, like that from a balloon and that he cannot, of course, expect to see them from the side, as he does the mountains upon the earth. But once he has realised this novel point of view, he will no doubt marvel at the formations which lie scattered as it were at his feet. The type of lunar mountain is indeed in striking contrast to the terrestrial type. On our earth the range-formation is supreme; on the moon the crater-formation is the rule, and is so-called from analogy to our volcanoes. A typical lunar crater may be described as a circular wall, enclosing a central plain, or "floor," which is often much depressed below the level of the surface outside. These so-called "craters," or "ring-mountains," as they are also termed, are often of gigantic proportions. For instance, the central plain of one of them, known as Ptolem鎬s,[17] is about 115 miles across, while that of Plato is about 60. The walls of craters often rise to great heights; which, in proportion to the small size of the moon, are very much in excess of our highest terrestrial elevations. Nevertheless, a person posted at the centre of one of the larger craters might be surprised to find that he could not see the encompassing crater-walls, which would in every direction be below his horizon. This would arise not alone from the great breadth of the crater itself, but also from the fact that the curving of the moon's surface is very sharp compared with that of our earth.

In this, as in the other plates of the Moon, the South will be found at the top of the picture; such being the view given by the ordinary astronomical telescope, in which all objects are seen inverted.

We have mentioned Ptolemes as among the very large craters, or ring-mountains, on the moon. Its encompassing walls rise to nearly 13,000 feet, and it has the further distinction of being almost in the centre of the lunar disc. There are, however, several others much wider, but they are by no means in such a conspicuous position. For instance, Schickard, close to the south-eastern border, is nearly 130 miles in diameter, and its wall rises in one point to over 10,000 feet. Grimaldi, almost exactly at the east point, is nearly as large as Schickard. Another crater, Clavius, situated near the south point, is about 140 miles across; while its neighbour Bailly--named after a famous French astronomer of the eighteenth century--is 180, and the largest of those which we can see (see Plate IX., p. 198).

Many of the lunar craters encroach upon one another; in fact there is not really room for them all upon the visible hemisphere of the moon. About 30,000 have been mapped; but this is only a small portion, for according to the American astronomer, Professor W.H. Pickering, there are more than 200,000 in all.

Notwithstanding the fact that the crater is the type of mountain associated in the mind with the moon, it must not be imagined that upon our satellite there are no mountains at all of the terrestrial type. There are indeed many isolated peaks, but strangely enough they are nearly always to be found in the centres of craters. Some of these peaks are of great altitude, that in the centre of the crater Copernicus being over 11,000 feet high. A few mountain ranges also exist; the best known of which are styled, the Lunar Alps and Lunar Apennines (see Plate X., p. 200).

Since the mass of the moon is only about one-eightieth that of the earth, it will be understood that the force of gravity which she exercises is much less. It is calculated that, at her surface, this is only about one-sixth of what we experience. A man transported to the moon would thus be able to jump six times as high as he can here. A building could therefore be six times as tall as upon our earth, without causing any more strain upon its foundations. It should not, then, be any subject for wonder, that the highest peaks in the Lunar Apennines attain to such heights as 22,000 feet. Such a height, upon a comparatively small body like the moon, for her volume is only one-fiftieth that of the earth, is relatively very much in excess of the 29,000 feet of Himalayan structure, Mount Everest, the boast of our planet, 8000 miles across!

High as are the Lunar Apennines, the highest peaks on the moon are yet not found among them. There is, for instance, on the extreme southern edge of the lunar disc, a range known as the Leibnitz Mountains; several peaks of which rise to a height of nearly 30,000 feet, one peak in particular being said to attain to 36,000 feet (see Plate IX., p. 198).

We have here (see "Map," Plate IX., p. 198) the mountain ranges of the Apennines, the Caucasus and the Alps; also the craters Plato, Aristotle, Eudoxus, Cassini, Aristillus, Autolycus, Archimedes and Linn? The crater Linn?is the very bright spot in the dark area at the upper left hand side of the picture. From a photograph taken at the Paris Observatory by M.M. Loewy and Puiseux.

(Page 200)]

But the reader will surely ask the question: "How is it possible to determine the actual height of a lunar mountain, if one cannot go upon the moon to measure it?" The answer is, that we can calculate its height from noting the length of the shadow which it casts. Any one will allow that the length of a shadow cast by the sun depends upon two things: firstly, upon the height of the object which causes the shadow, and secondly, upon the elevation of the sun at the moment in the sky. The most casual observer of nature upon our earth can scarcely have failed to notice that shadows are shortest at noonday, when the sun is at its highest in the sky; and that they lengthen out as the sun declines towards its setting. Here, then, we have the clue. To ascertain, therefore, the height of a lunar mountain, we have first to consider at what elevation the sun is at that moment above the horizon of the place where the mountain in question is situated. Then, having measured the actual length in miles of the shadow extended before us, all that is left is to ask ourselves the question: "What height must an object be whose shadow cast by the sun, when at that elevation in the sky, will extend to this length?"

There is no trace whatever of water upon the moon. The opinion, indeed, which seems generally held, is that water has never existed upon its surface. Erosions, sedimentary deposits, and all those marks which point to a former occupation by water are notably absent.

Similarly there appears to be no atmosphere on the moon; or, at any rate, such an excessively rare one, as to be quite inappreciable. Of this there are several proofs. For instance, in a solar eclipse the moon's disc always stands out quite clear-cut against that of the sun. Again during occultations, stars disappear behind the moon with a suddenness, which could not be the case were there any appreciable atmosphere. Lastly, we see no traces of twilight upon the lunar surface, nor any softening at the edges of shadows; both which effects would be apparent if there were an atmosphere.

The moon's surface is rough and rocky, and displays no marks of the "weathering" that would necessarily follow, had it possessed anything of an atmosphere in the past. This makes us rather inclined to doubt that it ever had one at all. Supposing, however, that it did possess an atmosphere in the past, it is interesting to inquire what may have become of it. In the first place it might have gradually disappeared, in consequence of the gases which composed it uniting chemically with the materials of which the lunar body is constructed; or, again, its constituent gases may have escaped into space, in accordance with the principles of that kinetic theory of which we have already spoken. The latter solution seems, indeed, the most reasonable of the two, for the force of gravity at the lunar surface appears too weak to hold down any known gases. This argument seems also to dispose of the question of absence of water; for Dr. George Johnstone Stoney, in a careful investigation of the subject, has shown that the liquid in question, when in the form of vapour, will escape from a planet if its mass is less than one-fourth that of our earth. And the mass of the moon is very much less than this; indeed only the one-eightieth, as we have already stated.

In consequence of this lack of atmosphere, the condition of things upon the moon will be in marked contrast to what we experience upon the earth. The atmosphere here performs a double service in shielding us from the direct rays of the sun, and in bottling the heat as a glass-house does. On the moon, however, the sun beats down in the day-time with a merciless force; but its rays are reflected away from the surface as quickly as they are received, and so the cold of the lunar night is excessive. It has been calculated that the day temperature on the moon may, indeed, be as high as our boiling-point, while the night temperature may be more than twice as low as the greatest cold known in our arctic regions.

That a certain amount of solar heat is reflected to us from the moon is shown by the sharp drop in temperature which certain heat-measuring instruments record when the moon becomes obscured in a lunar eclipse. The solar heat which is thus reflected to us by the moon is, however, on the whole extremely small; more light and heat, indeed, reach us direct from the sun in half a minute than we get by reflection from the moon during the entire course of the year.

With regard to the origin of the lunar craters there has been much discussion. Some have considered them to be evidence of violent volcanic action in the dim past; others, again, as the result of the impact of meteorites upon the lunar surface, when the moon was still in a plastic condition; while a third theory holds that they were formed by the bursting of huge bubbles during the escape into space of gases from the interior. The question is, indeed, a very difficult one. Though volcanic action, such as would result in craters of the size of Ptolem 鎢 s, is hard for us to picture, and though the lone peaks which adorn the centres of many craters have nothing reminiscent of them in our terrestrial volcanoes, nevertheless the volcanic theory seems to receive more favour than the others.

In addition to the craters there are two more features which demand notice, namely, what are known as rays and rills. The rays are long, light-coloured streaks which radiate from several of the large craters, and extend to a distance of some hundreds of miles. That they are mere markings on the surface is proved by the fact that they cast no shadows of any kind. One theory is, that they were originally great cracks which have been filled with lighter coloured material, welling up from beneath. The rills, on the other hand, are actually fissures, about a mile or so in width and about a quarter of a mile in depth.

The rays are seen to the best advantage in connection with the craters Tycho and Copernicus (see Plate XI., p. 204). In consequence of its fairly forward position on the lunar disc, and of the remarkable system of rays which issue from it like spokes from the axle of a wheel, Tycho commands especial attention. The late Rev. T.W. Webb, a famous observer, christened it, very happily, the "metropolitan crater of the moon."

The systems of rays from the craters Tycho, Copernicus and Kepler are well shown here. From a photograph taken at the Paris Observatory by M.P. Puiseux.

A great deal of attention is, and has been, paid by certain astronomers to the moon, in the hope of finding out if any changes are actually in progress at present upon her surface. Sir William Herschel, indeed, once thought that he saw a lunar volcano in eruption, but this proved to be merely the effect of the sunlight striking the top of the crater Aristarchus, while the region around it was still in shadow--sunrise upon Aristarchus, in fact! No change of any real importance has, however, been noted, although it is suspected that some minor alterations have from time to time taken place. For instance, slight variations of tint have been noticed in certain areas of the lunar surface. Professor W.H. Pickering puts forward the conjecture that these may be caused by the growth and decay of some low form of vegetation, brought into existence by vapours of water, or carbonic acid gas, making their way out from the interior through cracks near at hand.

Again, during the last hundred years one small crater known as Linn (Linneus), situated in the Mare Serenitatis (Sea of Serenity), has appeared to undergo slight changes, and is even said to have been invisible for a while (see Plate X., p. 200). It is, however, believed that the changes in question may be due to the varying angles at which the sunlight falls upon the crater; for it is an understood fact that the irregularities of the moon's motion give us views of her surface which always differ slightly.

The suggestion has more than once been put forward that the surface of the moon is covered with a thick layer of ice. This is generally considered improbable, and consequently the idea has received very little support. It first originated with the late Mr. S.E. Peal, an English observer of the moon, and has recently been resuscitated by the German observer, Herr Fauth.

The most unfavourable time for telescopic study of the moon is when she is full. The sunlight is then falling directly upon her visible hemisphere, and so the mountains cast no shadows. We thus do not get that impression of hill and hollow which is so very noticeable in the other phases.

The first map of the moon was constructed by Galileo. Tobias Mayer published another in 1775; while during the nineteenth century greatly improved ones were made by Beer and M鋭ler, Schmidt, Neison and others. In 1903, Professor W.H. Pickering brought out a complete photographic lunar atlas; and a similar publication has recently appeared, the work of MM. Loewy and Puiseux of the Observatory of Paris.

The so-called "seas" of the moon are, as we have seen, merely dark areas, and there appears to be no proof that they were ever occupied by any liquid. They are for the most part found in the northern portion of the moon; a striking contrast to our seas and oceans, which take up so much of the southern hemisphere of the earth.

There are many erroneous ideas popularly held with regard to certain influences which the moon is supposed to exercise upon the earth. For instance, a change in the weather is widely believed to depend upon a change in the moon. But the word "change" as here used is meaningless, for the moon is continually changing her phase during the whole of her monthly round. Besides, the moon is visible over a great portion of the earth at the same moment, and certainly all the places from which it can then be seen do not get the same weather! Further, careful observations, and records extending over the past one hundred years and more, fail to show any reliable connection between the phases of the moon and the condition of the weather.

It has been stated, on very good authority, that no telescope ever shows the surface of the moon as clearly as we could see it with the naked eye were it only 240 miles distant from us.

Supposing, then, that we were able to approach our satellite, and view it without optical aid at such comparatively close quarters, it is interesting to consider what would be the smallest detail which our eye could take in. The question of the limit of what can be appreciated with the naked eye is somewhat uncertain, but it appears safe to say that at a distance of 240 miles the minutest speck visible would have to be at least some 60 yards across.

Atmosphere and liquid both wanting, the lunar surface must be the seat of an eternal calm; where no sound breaks the stillness and where change, as we know it, does not exist. The sun beats down upon the arid rocks, and inky shadows lie athwart the valleys. There is no mellowing of the harsh contrasts.

We cannot indeed absolutely affirm that Life has no place at all upon this airless and waterless globe, since we know not under what strange conditions it may manifest its presence; and our most powerful telescopes, besides, do not bring the lunar surface sufficiently near to us to disprove the existence there of even such large creatures as disport themselves upon our planet. Still, we find it hard to rid ourselves of the feeling that we are in the presence of a dead world. On she swings around the earth month after month, with one face ever turned towards us, leaving a certain mystery to hang around that hidden side, the greater part of which men can never hope to see. The rotation of the moon upon her axis--the lunar day--has become, as we have seen, equal to her revolution around the earth. An epoch may likewise eventually be reached in the history of our own planet, when the length of the terrestrial day has been so slowed down by tidal friction that it will be equal to the year. Then will the earth revolve around the central orb, with one side plunged in eternal night and the other in eternal sunshine. But such a vista need not immediately distress us. It is millions of years forward in time.

[14] Journal of the British Astronomical Association, vol. x. (1899-1900), Nos. 1 and 3.

[15] Certain of the ancient Greeks thought the markings on the moon to be merely the reflection of the seas and lands of our earth, as in a badly polished mirror.

[16] Mare Imbrium, Sinus Iridum, Lacus Somniorum.

[17] The lunar craters have, as a rule, received their names from celebrated persons, usually men of science. This system of nomenclature was originated by Riccioli, in 1651.

CHAPTER XVII

THE SUPERIOR PLANETS

Having, in a previous chapter, noted the various aspects which an inferior planet presents to our view, in consequence of its orbit being nearer to the sun than the orbit of the earth, it will be well here to consider in the same way the case of a superior planet, and to mark carefully the difference.

To begin with, it should be quite evident that we cannot ever have a transit of a superior planet. The orbit of such a body being entirely outside that of the earth, the body itself can, of course, never pass between us and the sun.

A superior planet will be at its greatest distance from us when on the far side of the sun. It is said then to be in conjunction. As it comes round in its orbit it eventually passes, so to speak, at the back of us. It is then at its nearest, or in opposition, as this is technically termed, and therefore in the most favourable position for telescopic observation of its surface. Being, besides, seen by us at that time in the direction of the heavens exactly opposite to where the sun is, it will thus at midnight be high up in the south side of the sky, a further advantage to the observer.

Last of all, a superior planet cannot show crescent shapes like an interior; for whether it be on the far side of the sun, or behind us, or again to our right or left, the sunlight must needs appear to fall more or less full upon its face.

THE PLANETOID EROS

The nearest to us of the superior planets is the tiny body, Eros, which, as has been already stated, was discovered so late as the year 1898. In point of view, however, of its small size, it can hardly be considered as a true planet, and the name "planetoid" seems much more appropriate to it.

Eros was not discovered, like Uranus, in the course of telescopic examination of the heavens, nor yet, like Neptune, as the direct result of difficult calculations, but was revealed by the impress of its light upon a photographic plate, which had been exposed for some length of time to the starry sky. Since many of the more recent additions to the asteroids have been discovered in the same manner, we shall have somewhat more to say about this special employment of photography when we come to deal with those bodies later on.

The path of Eros around the sun is so very elliptical, or, to use the exact technical term, so very "eccentric," that the planetoid does not keep all the time entirely in the space between our orbit and that of Mars, which latter happens to be the next body in the order of planetary succession outwards. In portions of its journey Eros, indeed, actually goes outside the Martian orbit. The paths of the planetoid and of Mars are, however, not upon the same plane, so the bodies always pass clear of each other, and there is thus as little chance of their dashing together as there would be of trains which run across a bridge at an upper level, colliding with those which pass beneath it at a lower level.

When Eros is in opposition, it comes within about 13-1/2 million miles of our earth, and, after the moon, is therefore by a long way our nearest neighbour in space. It is, however, extremely small, not more, perhaps, than twenty miles in diameter, and is subject to marked variations in brightness, which do not appear up to the present to meet with a satisfactory explanation. But, insignificant as is this little body, it is of great importance to astronomy; for it happens to furnish the best method known of calculating the sun's distance from our earth--a method which Galle, in 1872, and Sir David Gill, in 1877, suggested that asteroids might be employed for, and which has in consequence supplanted the old one founded upon transits of Venus. The sun's distance is now an ascertained fact to within 100,000 miles, or less than half the distance of the moon.

THE PLANET MARS

We next come to the planet Mars. This body rotates in a period of slightly more than twenty-four hours. The inclination, or slant, of its axis is about the same as that of the earth, so that, putting aside its greater distance from the sun, the variations of season which it experiences ought to be very much like ours.

The first marking detected upon Mars was the notable one called the Syrtis Major, also known, on account of its shape, as the Hour-Glass Sea. This observation was made by the famous Huyghens in 1659; and, from the movement of the marking in question across the disc, he inferred that the planet rotated on its axis in a period of about twenty-four hours.

There appears to be very little atmosphere upon Mars, the result being that we almost always obtain a clear view of the detail on its surface. Indeed, it is only to be expected from the kinetic theory that Mars could not retain much of an atmosphere, as the force of gravity at its surface is less than one-half of what we experience upon the earth. It should here be mentioned that recent researches with the spectroscope seem to show that, whatever atmosphere there may be upon Mars, its density at the surface of the planet cannot be more than the one-fourth part of the density of the air at the surface of the earth. Professor Lowell, indeed, thinks it may be more rarefied than that upon our highest mountain-tops.

Seen with the naked eye, Mars appears of a red colour. Viewed in the telescope, its surface is found to be in general of a ruddy hue, varied here and there with darker patches of a bluish-green colour. These markings are permanent, and were supposed by the early telescopic observers to imply a distribution of the planet's surface into land and water, the ruddy portions being considered as continental areas (perhaps sandy deserts), and the bluish-green as seas. The similarity to our earth thus suggested was further heightened by the fact that broad white caps, situated at the poles, were seen to vary with the planet's seasons, diminishing greatly in extent during the Martian summer (the southern cap in 1894 even disappearing altogether), and developing again in the Martian winter.[18] Readers of Oliver Wendell Holmes will no doubt recollect that poet's striking lines:--

"The snows that glittered on the disc of Mars Have melted, and the planet's fiery orb Rolls in the crimson summer of its year."

A state of things so strongly analogous to what we experience here, naturally fired the imaginations of men, and caused them to look on Mars as a world like ours, only upon a much smaller scale. Being smaller, it was concluded to have cooled quicker, and to be now long past its prime; and its "inhabitants" were, therefore, pictured as at a later stage of development than the inhabitants of our earth.

Notwithstanding the strong temptation to assume that the whiteness of the Martian polar caps is due to fallen snow, such a solution is, however, by no means so simple as it looks. The deposition of water in the form of snow, or even of hoar frost, would at least imply that the atmosphere of Mars should now and then display traces of aqueous vapour, which it does not appear to do.[19] It has, indeed, been suggested that the whiteness may not after all be due to this cause, but to carbonic acid gas (carbon dioxide), which is known to freeze at a very low temperature. The suggestion is plainly based upon the assumption that, as Mars is so much further from the sun than we are, it would receive much less heat, and that the little thus received would be quickly radiated away into space through lack of atmosphere to bottle it in.

We now come to those well-known markings, popularly known as the "canals" of Mars, which have been the subject of so much discussion since their discovery thirty years ago.

It was, in fact, in the year 1877, when Mars was in opposition, and thus at its nearest to us, that the famous Italian astronomer, Schiaparelli, announced to the world that he had found that the ruddy areas, thought to be continents, were intersected by a network of straight dark lines. These lines, he reported, appeared in many cases to be of great length, so long, indeed, as several thousands of miles, and from about twenty to sixty miles in width. He christened the lines channels, the Italian word for which, "canali," was unfortunately translated into English as "canals." The analogy, thus accidentally suggested, gave rise to the idea that they might be actual waterways.[20]

In the winter of 1881-1882, when Mars was again in opposition, Schiaparelli further announced that he had found some of these lines doubled; that is to say, certain of them were accompanied by similar lines running exactly parallel at no great distance away. There was at first a good deal of scepticism on the subject of Schiaparelli's discoveries, but gradually other observers found themselves seeing both the lines and their doublings. We have in this a good example of a curious circumstance in astronomical observation, namely, the fact that when fine detail has once been noted by a competent observer, it is not long before other observers see the same detail with ease.

An immense amount of close attention has been paid to the planet Mars during recent years by the American observer, Professor Percival Lowell, at his famous observatory, 7300 feet above the sea, near the town of Flagstaff, Arizona, U.S.A. His observations have not, like those of most astronomers, been confined merely to "oppositions," but he has systematically kept the planet in view, so far as possible, since the year 1894.

The instrumental equipment of his observatory is of the very best, and the "seeing" at Flagstaff is described as excellent. In support of the latter statement, Mr. Lampland, of the Lowell Observatory, maintains that the faintest stars shown on charts made at the Lick Observatory with the 36-inch telescope there, are perfectly visible with the 24-inch telescope at Flagstaff.

Professor Lowell is, indeed, generally at issue with the other observers of Mars. He finds the canals extremely narrow and sharply defined, and he attributes the blurred and hazy appearance, which they have presented to other astronomers, to the unsteady and imperfect atmospheric conditions in which their observations have been made. He assigns to the thinnest a width of two or three miles, and from fifteen to twenty to the larger. Relatively to their width, however, he finds their length enormous. Many of them are 2000 miles long, while one is even as much as 3540. Such lengths as these are very great in comparison with the smallness of the planet. He considers that the canals stand in some peculiar relation to the polar cap, for they crowd together in its neighbourhood. In place, too, of ill-defined condensations, he sees sharp black spots where the canals meet and intersect, and to these he gives the name of "Oases." He further lays particular stress upon a dark band of a blue tint, which is always seen closely to surround the edges of the polar caps all the time that they are disappearing; and this he takes to be a proof that the white material is something which actually melts. Of all substances which we know, water alone, he affirms, would act in such a manner.

The question of melting at all may seem strange in a planet which is situated so far from the sun, and possesses such a rarefied atmosphere. But Professor Lowell considers that this very thinness of the atmosphere allows the direct solar rays to fall with great intensity upon the planet's surface, and that this heating effect is accentuated by the great length of the Martian summer. In consequence he concludes that, although the general climate of Mars is decidedly cold, it is above the freezing point of water.

The observations at Flagstaff appear to do away with the old idea that the darkish areas are seas, for numerous lines belonging to the so-called "canal system" are seen to traverse them. Again, there is no star-like image of the sun reflected from them, as there would be, of course, from the surface of a great sheet of water. Lastly, they are observed to vary in tone and colour with the changing Martian seasons, the blue-green changing into ochre, and later on back again into blue-green. Professor Lowell regards these areas as great tracts of vegetation, which are brought into activity as the liquid reaches them from the melting snows.

We see here the Syrtis Major (or "Hour-Glass Sea"), the polar caps, several "oases," and a large number of "canals," some of which are double. The South is at the top of the picture, in accordance with the inverted view given by an astronomical telescope. From a drawing by Professor Percival Lowell.

With respect to the canals, the Lowell observations further inform us that these are invisible during the Martian winter, but begin to appear in the spring when the polar cap is disappearing. Professor Lowell, therefore, inclines to the view that in the middle of the so-called canals there exist actual waterways which serve the purposes of irrigation, and that what we see is not the waterways themselves, for they are too narrow, but the fringe of vegetation which springs up along the banks as the liquid is borne through them from the melting of the polar snows. He supports this by his observation that the canals begin to appear in the neighbourhood of the polar caps, and gradually grow, as it were, in the direction of the planet's equator.

It is the idea of life on Mars which has given this planet such a fascination in the eyes of men. A great deal of nonsense has, however, been written in newspapers upon the subject, and many persons have thus been led to think that we have obtained some actual evidence of the existence of living beings upon Mars. It must be clearly understood, however, that Professor Lowell's advocacy of the existence of life upon that planet is by no means of this wild order. At the best he merely indulges in such theories as his remarkable observations naturally call forth. His views are as follows:--He considers that the planet has reached a time when "water" has become so scarce that the "inhabitants" are obliged to employ their utmost skill to make their scanty supply suffice for purposes of irrigation. The changes of tone and colour upon the Martian surface, as the irrigation produces its effects, are similar to what a telescopic observer--say, upon Venus--would notice on our earth when the harvest ripens over huge tracts of country; that is, of course, if the earth's atmosphere allowed a clear view of the terrestrial surface--a very doubtful point indeed. Professor Lowell thinks that the perfect straightness of the lines, and the geometrical manner in which they are arranged, are clear evidences of artificiality. On a globe, too, there is plainly no reason why the liquid which results from the melting of the polar caps should trend at all in the direction of the equator. Upon our earth, for instance, the transference of water, as in rivers, merely follows the slope of the ground, and nothing else. The Lowell observations show, however, that the Martian liquid is apparently carried from one pole towards the equator, and then past it to the other pole, where it once more freezes, only to melt again in due season, and to reverse the process towards and across the equator as before. Professor Lowell therefore holds, and it seems a strong point in favour of his theory, that the liquid must, in some artificial manner, as by pumping, for instance, be helped in its passage across the surface of the planet.

A number of attempts have been made to explain the doubling of the canals merely as effects of refraction or reflection; and it has even been suggested that it may arise from the telescope not being accurately focussed.

The actual doubling of the canals once having been doubted, it was an easy step to the casting of doubt on the reality of the canals themselves. The idea, indeed, was put forward that the human eye, in dealing with detail so very close to the limit of visibility, may unconsciously treat as an actual line several point-like markings which merely happen to lie in a line. In order to test this theory, experiments were carried out in 1902 by Mr. E.W. Maunder of

Greenwich Observatory, and Mr. J.E. Evans of the Royal Hospital School at Greenwich, in which certain schoolboys were set to make drawings of a white disc with some faint markings upon it. The boys were placed at various distances from the disc in question; and it was found that the drawings made by those who were just too far off to see distinctly, bore out the above theory in a remarkable manner. Recently, however, the plausibility of the illusion view has been shaken by photographs of Mars taken during the opposition of 1905 by Mr. Lampland at the Lowell Observatory, in which a number of the more prominent canals come out as straight dark lines. Further still, in some photographs made there quite lately, several canals are said to appear visibly double.

Following up the idea alluded to in Chapter XVI., that the moon may be covered with a layer of ice, Mr. W.T. Lynn has recently suggested that this may be the case on Mars; and that, at certain seasons, the water may break through along definite lines, and even along lines parallel to these. This, he maintains, would account for the canals becoming gradually visible across the disc, without the necessity of Professor Lowell's "pumping" theory.

And now for the views of Professor Lowell himself with regard to the doubling of the canals. From his observations, he considers that no pairs of railway lines could apparently be laid down with greater parallelism. He draws attention to the fact that the doubling does not take place by any means in every canal; indeed, out of 400 canals seen at Flagstaff, only fifty-one--or, roughly, one-eighth--have at any time been seen double. He lays great stress upon this, which he considers points strongly against the duplication being an optical phenomenon. He finds that the distance separating pairs of canals is much less in some doubles than in others, and varies on the whole from 75 to 200 miles. According to him, the double canals appear to be confined to within 40 degrees of the equator: or, to quote his own words, they are "an equatorial feature of the planet, confined to the tropic and temperate belts." Finally, he points out that they seem to avoid the blue-green areas. But, strangely enough, Professor Lowell does not so far attempt to fit in the doubling with his body of theory. He makes the obvious remark that they may be "channels and return channels," and with that he leaves us.

The conclusions of Professor Lowell have recently been subjected to strenuous criticism by Professor W.H. Pickering and Dr. Alfred Russel Wallace. It was Professor Pickering who discovered the "oases," and who originated the idea that we did not see the so-called "canals" themselves, but only the growth of vegetation along their borders. He holds that the oases are craterlets, and that the canals are cracks which radiate from them, as do the rifts and streaks from craters upon the moon. He goes on to suggest that vapours of water, or of carbonic acid gas, escaping from the interior, find their way out through these cracks, and promote the growth of a low form of vegetation on either side of them. In support of this view he draws attention to the existence of long "steam-cracks," bordered by vegetation, in the deserts of the highly volcanic island of Hawaii. We have already seen, in an earlier chapter, how he has applied this idea to the explanation of certain changes which are suspected to be taking place upon the moon.

In dealing with the Lowell canal system, Professor Pickering points out that under such a slight atmospheric pressure as exists on Mars, the evaporation of the polar caps--supposing them to be formed of snow--would take place with such extraordinary rapidity that the resulting water could never be made to travel along open channels, but that a system of gigantic tubes or water-mains would have to be employed!

As will be gathered from his theories regarding vegetation, Professor Pickering does not deny the existence of a form of life upon Mars. But he will not hear of civilisation, or of anything even approaching it. He thinks, however, that as Mars is intermediate physically between the moon and earth, the form of life which it supports may be higher than that on the moon and lower than that on the earth.

In a small book published in the latter part of 1907, and entitled Is Mars Habitable? Dr. Alfred Russel Wallace sets himself, among other things, to combat the idea of a comparatively high temperature, such as Professor Lowell has allotted to Mars. He shows the immense service which the water-vapour in our atmosphere exercises, through keeping the solar heat from escaping from the earth's surface. He then draws attention to the fact that there is no spectroscopic evidence of water-vapour on Mars[21]; and points out that its absence is only to be expected, as Dr. George Johnstone Stoney has shown that it will escape from a body whose mass is less than one-quarter the mass of the earth. The mass of Mars is, in fact, much less than this, i.e. only one-ninth. Dr. Wallace considers, therefore, that the temperature of Mars ought to be extremely low, unless the constitution of its atmosphere is very different from ours. With regard to the latter statement, it should be mentioned that the Swedish physicist, Arrhenius, has recently shown that the carbonic acid gas in our atmosphere has an important influence upon climate. The amount of it in our air is, as we have seen, extremely small; but Arrhenius shows that, if it were doubled, the temperature would be more uniform and much higher. We thus see how futile it is, with our present knowledge, to dogmatise on the existence or non-existence of life in other celestial orbs.

As to the canals Dr. Wallace puts forward a theory of his own. He contends that after Mars had cooled to a state of solidity, a great swarm of meteorites and small asteroids fell in upon it, with the result that a thin molten layer was formed all over the planet. As this layer cooled, the imprisoned gases escaped, producing vents or craterlets; and as it attempted to contract further upon the solid interior, it split in fissures radiating from points of weakness, such, for instance, as the craterlets. And he goes on to suggest that the two tiny Martian satellites, with which we shall deal next, are the last survivors of his hypothetical swarm. Finally, with regard to the habitability of Mars, Dr. Wallace not only denies it, but asserts that the planet is "absolutely uninhabitable."

For a long time it was supposed that Mars did not possess any satellites. In 1877, however, during that famous opposition in which Schiaparelli first saw the canals, two tiny satellites were discovered at the Washington Observatory by an American astronomer, Professor Asaph Hall. These satellites are so minute, and so near to the planet, that they can only be seen with very large telescopes; and even then the bright disc of the planet must be shielded off. They have been christened Phobos and Deimos (Fear and Dread); these being the names of the two subordinate deities who, according to Homer, attended upon Mars, the god of war.

It is impossible to measure the exact sizes of these satellites, as they are too small to show any discs, but an estimate has been formed from their brightness. The diameter of Phobos was at first thought to be six miles, and that of Deimos, seven. As later estimates, however, considerably exceed this, it will, perhaps, be not far from the truth to state that they are each roughly about the size of the planetoid Eros. Phobos revolves around Mars in about 7-1/2 hours, at a distance of about only 4000 miles from the planet's surface, and Deimos in about 30 hours, at a distance of about 12,000 miles. As Mars rotates on its axis in about 24 hours, it will be seen that Phobos makes more than three revolutions while the planet is rotating once--a very interesting condition of things.

A strange foreshadowing of the discovery of the satellites of Mars will be familiar to readers of Gulliver's Travels. According to Dean Swift's hero, the astronomers on the Flying Island of Laputa had found two tiny satellites to Mars, one of which revolved around the planet in ten hours. The correctness of this guess is extraordinarily close, though at best it is, of course, nothing more than a pure coincidence.

It need not be at all surprising that much uncertainty should exist with regard to the actual condition of the surface of Mars. The circumstances in which we are able to see that planet at the best are, indeed, hardly sufficient to warrant us in propounding any hard and fast theories. One of the most experienced of living observers, the American astronomer, Professor E.E. Barnard, considers that the view we get of Mars with the best telescope may be fairly compared with our naked eye view of the moon. Since we have seen that a view with quite a small telescope entirely alters our original idea of the lunar surface, a slight magnification revealing features of whose existence we had not previously the slightest conception, it does not seem too much to say that a further improvement in optical power might entirely subvert the present notions with regard to the Martian canals. Therefore, until we get a still nearer view of these strange markings, it seems somewhat futile to theorise. The lines which we see are perhaps, indeed, a foreshortened and all too dim view of some type of formation entirely novel to us, and possibly peculiar to Mars. Differences of gravity and other conditions, such as obtain upon different planets, may perhaps produce very diverse results. The earth, the moon, and Mars differ greatly from one another in size, gravitation, and other such characteristics. Mountain-ranges so far appear typical of our globe, and ring-mountains typical of the moon. May not the so-called "canals" be merely some special formation peculiar to Mars, though quite a natural result of its particular conditions and of its past history?

THE ASTEROIDS (OR MINOR PLANETS)

We now come to that belt of small planets which are known by the name of asteroids. In the general survey of the solar system given in Chapter II., we saw how it was long ago noticed that the distances of the planetary orbits from the sun would have presented a marked appearance of orderly sequence, were it not for a gap between the orbits of Mars and Jupiter where no large planet was known to circulate. The suspicion thus aroused that some planet might, after all, be moving in this seemingly empty space, gave rise to the gradual discovery of a great number of small bodies; the largest of which, Ceres, is less than 500 miles in diameter. Up to the present day some 600 of these bodies have been discovered; the four leading ones, in order of size, being named Ceres, Pallas, Juno, and Vesta. All the asteroids are invisible to the naked eye, with the exception of Vesta, which, though by no means the largest, happens to be the brightest. It is, however, only just visible to the eye under favourable conditions. No trace of an atmosphere has been noted upon any of the asteroids, but such a state of things is only to be expected from the kinetic theory.

For a good many years the discoveries of asteroids were made by means of the telescope. When, in the course of searching the heavens, an object was noticed which did not appear upon any of the recognised star charts, it was kept under observation for several nights to see whether it changed its place in the sky. Since asteroids move around the sun in orbits, just as planets do, they, of course, quickly reveal themselves by their change of position against the starry background.

The year 1891 started a new era in the discovery of asteroids. It occurred to the Heidelberg observer, Dr. Max Wolf, one of the most famous of the hunters of these tiny planets, that photography might be employed in the quest with success. This photographic method, to which allusion has already been made in dealing with Eros, is an extremely simple one. If a photograph of a portion of the heavens be taken through an "equatorial"--that is, a telescope, moving by machinery, so as to keep the stars, at which it is pointed, always exactly in the field of view during their apparent movement across the sky--the images of these stars will naturally come out in the photograph as sharply defined points. If, however, there happens to be an asteroid, or other planetary body, in the same field of view, its image will come out as a short white streak; because the body has a comparatively rapid motion of its own, and will, during the period of exposure, have moved sufficiently against the background of the stars to leave a short trail, instead of a dot, upon the photographic plate. By this method Wolf himself has succeeded in discovering more than a hundred asteroids (see Plate XIII., p. 226). It was, indeed, a little streak of this kind, appearing upon a photograph taken by the astronomer Witt, at Berlin, in 1898, which first informed the world of the existence of Eros.

Two trails of minor planets (asteroids) imprinted at the same time upon one photographic plate. In the white streak on the left-hand side of the picture we witness the discovery of a new minor planet. The streak on the right was made by a body already known--the minor planet "Fiducia." This photograph was taken by Dr. Max Wolf, at Heidelberg, on the 4th of November, 1901, with the aid of a 16-inch telescope. The time of exposure was two hours.

It has been calculated that the total mass of the asteroids must be much less than one-quarter that of the earth. They circulate as a rule within a space of some 30,000,000 miles in breadth, lying about midway between the paths of Mars and Jupiter. Two or three, however, of the most recently discovered of these small bodies have been found to pass quite close to Jupiter. The orbits of the asteroids are by no means in the one plane, that of Pallas being the most inclined to the plane of the earth's orbit. It is actually three times as much inclined as that of Eros.

Two notable theories have been put forward to account for the origin of the asteroids. The first is that of the celebrated German astronomer, Olbers, who was the discoverer of Pallas and Vesta. He suggested that they were the fragments of an exploded planet. This theory was for a time generally accepted, but has now been abandoned in consequence of certain definite objections. The most important of these objections is that, in accordance with the theory of gravitation, the orbits of such fragments would all have to pass through the place where the explosion originally occurred. But the wide area over which the asteroids are spread points rather against the notion that they all set out originally from one particular spot. Another objection is that it does not appear possible that, within a planet already formed, forces could originate sufficiently powerful to tear the body asunder.

The second theory is that for some reason a planet here failed in the making.

Possibly the powerful gravitational action of the huge body of Jupiter hard by, disturbed this region so much that the matter distributed through it was never able to collect itself into a single mass.

[18] Sir William Herschel was the first to note these polar changes.

[19] Quite recently, however, Professor Lowell has announced that his observer, Mr. E.C. Slipher, finds with the spectroscope faint traces of water vapour in the Martian atmosphere.

[20] In a somewhat similar manner the term "crater," as applied to the ring-mountain formation on the moon, has evidently given a bias in favour of the volcanic theory as an explanation of that peculiar structure.

[21] Mr. Slipher's results (see note 2, page 213) were not then known.

CHAPTER XVIII

THE SUPERIOR PLANETS--continued

The planets, so far, have been divided into inferior and superior. Such a division, however, refers merely to the situation of their orbits with regard to that of our earth. There is, indeed, another manner in which they are often classed, namely, according to size. On this principle they are divided into two groups; one group called the Terrestrial Planets, or those which have characteristics like our earth, and the other called the Major Planets, because they are all of very great size. The terrestrial planets are Mercury, Venus, the earth, and Mars. The major planets are the remainder, namely, Jupiter, Saturn, Uranus, and Neptune. As the earth's orbit is the boundary which separates the inferior from the superior planets, so does the asteroidal belt divide the terrestrial from the major planets. We found the division into inferior and superior useful for emphasising the marked difference in aspect which those two classes present as seen from our earth; the inferior planets showing phases like the moon when viewed in the telescope, whereas the superior planets do not. But the division into terrestrial and major planets is the more far-reaching classification of the two, for it includes the whole number of planets, whereas the other arrangement necessarily excludes the earth. The members of each of these classes have many definite characteristics in common. The terrestrial planets are all of them relatively small in size, comparatively near together, and have few or no satellites. They are, moreover, rather dense in structure. The major planets, on the other hand, are huge bodies, circulating at great distances from each other, and are, as a rule, provided with a number of satellites. With respect to structure, they may be fairly described as being loosely put together. Further, the markings on the surfaces of the terrestrial planets are permanent, whereas those on the major planets are continually shifting.

THE PLANET JUPITER

Jupiter is the greatest of the major planets. It has been justly called the "Giant" planet, for both in volume and in mass it exceeds all the other planets put together. When seen through the telescope it exhibits a surface plentifully covered with markings, the most remarkable being a series of broad parallel belts. The chief belt lies in the central parts of the planet, and is at present about 10,000 miles wide. It is bounded on either side by a reddish brown belt of about the same width. Bright spots also appear upon the surface of the planet, last for a while, and then disappear. The most notable of the latter is one known as the "Great Red Spot." This is situated a little beneath the southern red belt, and appeared for the first time about thirty years ago. It has undergone a good many changes in colour and brightness, and is still faintly visible. This spot is the most permanent marking which has yet been seen upon Jupiter. In general, the markings change so often that the surface which we see is evidently not solid, but of a fleeting nature akin to cloud

The Giant Planet as seen at 11.30 p.m., on the 11th of January, 1908, with a 12-1/2-inch reflecting telescope. The extensive oval marking in the upper portion of the disc is the "Great Red Spot." The South is at the top of the picture, the view being the inverted one given by an astronomical telescope. From a drawing by the Rev. Theodore E.R. Phillips, M.A., F.R.A.S., Director of the Jupiter Section of the British Astronomical Association.

Observations of Jupiter's markings show that on an average the planet rotates on its axis in a period of about 9 hours 54 minutes. The mention here of an average with reference to the rotation will, no doubt, recall to the reader's mind the similar case of the sun, the different portions of which rotate with different velocities. The parts of Jupiter which move quickest take 9 hours 50 minutes to go round, while those which move slowest take 9 hours 57 minutes. The middle portions rotate the fastest, a phenomenon which the reader will recollect was also the case with regard to the sun.

Jupiter is a very loosely packed body. Its density is on an average only about 1-1/2 times that of water, or about one-fourth the density of the earth; but its bulk is so great that the gravitation at that surface which we see is about 2-1/2 times what it is on the surface of the earth. In accordance, therefore, with the kinetic theory, we may expect the planet to retain an extensive layer of gases around it; and this is confirmed by the spectroscope, which gives evidence of the presence of a dense atmosphere.

All things considered, it may be safely inferred that the interior of Jupiter is very hot, and that what we call its surface is not the actual body of the planet, but a voluminous layer of clouds and vapours driven upwards from the heated mass underneath. The planet was indeed formerly thought to be self-luminous; but this can hardly be the case, for those portions of the surface which happen to lie at any moment in the shadows cast by the satellites appear to be quite black. Again, when a satellite passes into the great shadow cast by the planet it becomes entirely invisible, which would not be the case did the planet emit any perceptible light of its own.

In its revolutions around the sun, Jupiter is attended, so far as we know, by seven[22] satellites. Four of these were among the first celestial objects which Galileo discovered with his "optick tube," and he named them the "Medicean Stars" in honour of his patron, Cosmo de Medici. Being comparatively large bodies they might indeed just be seen with the naked eye, were it not for the overpowering glare of the planet.

It was only in quite recent times, namely, in 1892, that a fifth satellite was added to the system of Jupiter. This body, discovered by Professor E.E. Barnard, is very small. It circulates nearer to the planet than the innermost of Galileo's moons; and, on account of the glare, is a most difficult object to obtain a glimpse of, even in the best of telescopes. In December 1904 and January 1905 respectively, two more moons were added to the system, these being found by photography, by the American astronomer, Professor C.D. Perrine. Both the bodies in question revolve at a greater distance from the planet than the outermost of the older known satellites.

Galileo's moons, though the largest bodies of Jupiter's satellite system, are, as we have already pointed out, very small indeed when compared with the planet itself. The diameters of two of them, Europa and Io, are, however, about the same as that of our moon, while those of the other two, Callisto and Ganymede, are more than half as large again. The recently discovered satellites are, on the other hand, insignificant; that found by Barnard, for example, being only about 100 miles in diameter.

Of the four original satellites Io is the nearest to Jupiter, and, seen from the planet, it would show a disc somewhat larger than that of our moon. The others would appear somewhat smaller. However, on account of the great distance of the sun, the entire light reflected to Jupiter by all the satellites should be very much less than what we get from our moon.

Barnard's satellite circles around Jupiter at a distance less than our moon is from us, and in a period of about 12 hours. Galileo's four satellites revolve in periods of about 2, 3-1/2, 7, and 16-1/2 days respectively, at distances lying roughly between a quarter of a million and one million miles. Perrine's two satellites are at a distance of about seven million miles, and take about nine months to complete their revolutions.

The larger satellites, when viewed in the telescope, exhibit certain defined markings; but the bodies are so far away from us, that only those details which are of great extent can be seen. The satellite Io, according to Professor Barnard, shows a darkish disc, with a broad white belt across its middle regions. Mr. Douglass, one of the observers at the Lowell Observatory, has noted upon Ganymede a number of markings somewhat resembling those seen on Mars, and he concludes, from their movement, that this satellite rotates on its axis in about seven days. Professor Barnard, on the other hand, does not corroborate this, though he claims to have discovered bright polar caps on both Ganymede and Callisto.

In an earlier chapter we dealt at length with eclipses, occultations, and transits, and endeavoured to make clear the distinction between them. The system of Jupiter's satellites furnishes excellent examples of all these phenomena. The planet casts a very extensive shadow, and the satellites are constantly undergoing obscuration by passing through it. Such occurrences are plainly comparable to our lunar eclipses. Again, the satellites may, at one time, be occulted by the huge disc of the planet, and at another time seen in transit over its face. A fourth phenomenon is what is known as an eclipse of the planet by a satellite, which is the exact equivalent of what we style on the earth an eclipse of the sun. In this last case the shadow, cast by the satellite, appears as a round black spot in movement across the planet's surface.

In the passages of these attendant bodies behind the planet, into its shadow, or across its face, respectively, it occasionally happens that Galileo's four satellites all disappear from view, and the planet is then seen for a while in the unusual condition of being apparently without its customary attendants. An instance of this phenomenon took place on the 3rd of October 1907. On that occasion, the satellites known as I. and III. (i.e. Io and Ganymede) were eclipsed, that is to say, obscured by passing into the planet's shadow; Satellite IV. (Callisto) was occulted by the planet's disc; while Satellite II. (Europa), being at the same moment in transit across the planet's face, was invisible against that brilliant background. A number of instances of this kind of occurrence are on record. Galileo, for example, noted one on the 15th of March 1611, while Herschel observed another on the 23rd of May 1802.

It was indirectly to Jupiter's satellites that the world was first indebted for its knowledge of the velocity of light. When the periods of revolution of the satellites were originally determined, Jupiter happened, at the time, to be at his nearest to us. From the periods thus found tables were made for the prediction of the moments at which the eclipses and other phenomena of the satellites should take place. As Jupiter, in the course of his orbit, drew further away from the earth, it was noticed that the disappearances of the satellites into the shadow of the planet occurred regularly later than the time predicted. In the year 1675, Roemer, a Danish astronomer, inferred from this, not that the predictions were faulty, but that light did not travel instantaneously. It appeared, in fact, to take longer to reach us, the greater the distance it had to traverse. Thus, when the planet was far from the earth, the last ray given out by the satellite, before its passage into the shadow, took a longer time to cross the intervening space, than when the planet was near. Modern experiments in physics have quite confirmed this, and have proved for us that light does not travel across space in the twinkling of an eye, as might hastily be supposed, but actually moves, as has been already stated, at the rate of about 186,000 miles per second.

THE PLANET SATURN

Seen in the telescope the planet Saturn is a wonderful and very beautiful object. It is distinguished from all the other planets, in fact from all known celestial bodies, through being girt around its equator by what looks like a broad, flat ring of exceeding thinness. This, however, upon closer examination, is found to be actually composed of three concentric rings. The outermost of these is nearly of the same brightness as the body of the planet itself. The ring which comes immediately within it is also bright, and is separated from the outer one all the way round by a relatively narrow space, known as "Cassini's division," because it was discovered by the celebrated French astronomer, J.D. Cassini, in the year 1675. Inside the second ring, and merging insensibly into it, is a third one, known as the "crape ring," because it is darker in hue than the others and partly transparent, the body of Saturn being visible through it. The inner boundary of this third and last ring does not adjoin the planet, but is everywhere separated from it by a definite space. This ring was discovered independently[23] in 1850 by Bond in America and Dawes in England.

From a drawing made by Professor Barnard with the Great Lick Telescope. The black band fringing the outer ring, where it crosses the disc, is portion of the shadow which the rings cast upon the planet. The black wedge-shaped mark, where the rings disappear behind the disc at the left-hand side, is portion of the shadow which the planet casts upon the rings.

As distinguished from the crape ring, the bright rings must have a considerable closeness of texture; for the shadow of the planet may be seen projected upon them, and their shadows in turn projected upon the surface of the planet (see Plate XV., p. 236).

According to Professor Barnard, the entire breadth of the ring system, that is to say, from one side to the other of the outer ring, is 172,310 miles, or somewhat more than double the planet's diameter.

In the varying views which we get of Saturn, the system of the rings is presented to us at very different angles. Sometimes we are enabled to gaze upon its broad expanse; at other times, however, its thin edge is turned exactly towards us, an occurrence which takes place after intervals of about fifteen years. When this happened in 1892 the rings are said to have disappeared entirely from view in the great Lick telescope. We thus get an idea of their small degree of thickness, which would appear to be only about 50 miles. The last time the system of rings was exactly edgewise to the earth was on the 3rd of October 1907.

The question of the composition of these rings has given rise to a good deal of speculation. It was formerly supposed that they were either solid or liquid, but in 1857 it was proved by Clerk Maxwell that a structure of this kind would not be able to stand. He showed, however, that they could be fully explained by supposing them to consist of an immense number of separate solid particles, or, as one might otherwise put it, extremely small satellites, circling in dense swarms around the middle portions of the planet. It is therefore believed that we have here the materials ready for the formation of a satellite or satellites; but that the powerful gravitative action, arising through the planet's being so near at hand, is too great ever to allow these materials to aggregate themselves into a solid mass. There is, as a matter of fact, a minimum distance from the body of any planet within which it can be shown that a satellite will be unable to form on account of gravitational stress. This is known as "Roche's limit," from the name of a French astronomer who specially investigated the question.

There thus appears to be a certain degree of analogy between Saturn's rings and the asteroids. Empty spaces, too, exist in the asteroidal zone, the relative position of one of which bears a striking resemblance to that of "Cassini's division." It is suggested, indeed, that this division had its origin in gravitational disturbances produced by the attraction of the larger satellites, just as the empty spaces in the asteroidal zone are supposed to be the result of perturbations caused by the Giant Planet hard by.

It has long been understood that the system of the rings must be rotating around Saturn, for if they were not in motion his intense gravitational attraction would quickly tear them in pieces. This was at length proved to be the fact by the late Professor Keeler, Director of the Lick Observatory, who from spectroscopic observations found that those portions of the rings situated near to the planet rotated faster than those farther from it. This directly supports the view that the rings are composed of satellites; for, as we have already seen, the nearer a satellite is to its primary the faster it will revolve. On the other hand, were the rings solid, their outer portions would move the fastest; as we have seen takes place in the body of the earth, for example. The mass of the ring system, however, must be exceedingly small, for it does not appear to produce any disturbances in the movements of Saturn's satellites. From the kinetic theory, therefore, one would not expect to find any atmosphere on the rings, and the absence of it is duly shown by spectroscopic observations.

The diameter of Saturn, roughly speaking, is about one-fifth less than that of Jupiter. The planet is very flattened at the poles, this flattening being quite noticeable in a good telescope. For instance, the diameter across the equator is about 76,470 miles, while from pole to pole it is much less, namely, 69,770.

The surface of Saturn bears a strong resemblance to that of Jupiter. Its markings, though not so well defined, are of the same belt-like description; and from observation of them it appears that the planet rotates on an average in a little over ten hours. The rotation is in fact of the same peculiar kind as that of the sun and Jupiter; but the difference of speed at which the various portions of Saturn go round are even more marked than in the case of the Giant Planet. The density of Saturn is less than that of Jupiter; so that it must be largely in a condition of vapour, and in all probability at a still earlier stage of planetary evolution.

Up to the present we know of as many as ten satellites circling around Saturn, which is more than any other planet of the solar system can lay claim to. Two of these, however, are very recent discoveries; one, Phoebe, having been found by photography in August 1898, and the other, Themis, in 1904, also by the same means. For both of these we are indebted to Professor W.H. Pickering. Themis is said to be the faintest object in the solar system. It cannot be seen, even with the largest telescope in existence; a fact which should hardly fail to impress upon one the great advantage the photographic plate possesses in these researches over the human eye.

The most important of the whole Saturnian family of satellites are the two known as Titan and Japetus. These were discovered respectively by Huyghens in 1655 and by Cassini in 1671. Japetus is about the same size as our moon;

while the diameter of Titan, the largest of the satellites, is about half as much again. Titan takes about sixteen days to revolve around Saturn, while Japetus takes more than two months and a half. The former is about three-quarters of a million miles distant from the planet, and the latter about two and a quarter millions. To Sir William Herschel we are indebted for the discovery of two more satellites, one of which he found on the evening that he used his celebrated 40-foot telescope for the first time. The ninth satellite, Phoebe, one of the two discovered by Professor Pickering, is perhaps the most remarkable body in the solar system, for all the other known members of that system perform their revolutions in one fixed direction, whereas this satellite revolves in the contrary direction.

In consequence of the great distance of Saturn, the sun, as seen from the planet, would appear so small that it would scarcely show any disc. The planet, indeed, only receives from the sun about one-ninetieth of the heat and light which the earth receives. Owing to this diminished intensity of illumination, the combined light reflected to Saturn by the whole of its satellites must be very small.

With the sole exception of Jupiter, not one of the planets circulating nearer to the sun could be seen from Saturn, as they would be entirely lost in the solar glare. For an observer upon Saturn, Jupiter would, therefore, fill much the same position as Venus does for us, regularly displaying phases and being alternately a morning and an evening star.

It is rather interesting to consider the appearances which would be produced in our skies were the earth embellished with a system of rings similar to those of Saturn. In consequence of the curving of the terrestrial surface, they would not be seen at all from within the Arctic or Antarctic circles, as they would be always below the horizon. From the equator they would be continually seen edgewise, and so would appear merely as line of light stretching right across the heaven and passing through the zenith. But the dwellers in the remaining regions would find them very objectionable, for they would cut off the light of the sun during lengthy periods of time.

Saturn was a sore puzzle to the early telescopic observers. They did not for a long time grasp the fact that it was surrounded by a ring--so slow is the human mind to seek for explanations out of the ordinary course of things.

The protrusions of the ring on either side of the planet, at first looked to Galileo like two minor globes placed on opposite sides of it, and slightly overlapping the disc. He therefore informed Kepler that "Saturn consists of three stars in contact with one another." Yet he was genuinely puzzled by the fact that the two attendant bodies (as he thought them) always retained the same position with regard to the planet's disc, and did not appear to revolve around it, nor to be in any wise shifted as a consequence of the movements of our earth.

About a year and a half elapsed before he again examined Saturn; and, if he was previously puzzled, he was now thoroughly amazed. It happened just then to be one of those periods when the ring is edgewise towards the earth, and of course he only saw a round disc like that of Jupiter. What, indeed, had become of the attendant orbs? Was some demon mocking him? Had Saturn devoured his own children? He was, however, fated to be still more puzzled, for soon the minor orbs reappeared, and, becoming larger and larger as time went on, they ended by losing their globular appearance and became like two pairs of arms clasping the planet from each side! (see Plate XVI., p. 242).

Galileo went to his grave with the riddle still unsolved, and it remained for the famous Dutch astronomer, Huyghens, to clear up the matter. It was, however, some little time before he hit upon the real explanation. Having noticed that there were dark spaces between the strange appendages and the body of the planet, he imagined Saturn to be a globe fitted with handles at each side; "ans? these came to be called, from the Latin ansa, which means a handle. At length, in the year 1656, he solved the problem, and this he did by means of that 123-foot tubeless telescope, of which mention has already been made. The ring happened then to be at its edgewise period, and a careful study of the behaviour of the ans?when disappearing and reappearing soon revealed to Huyghens the true explanation.

THE PLANETS URANUS AND NEPTUNE

We have already explained (in Chapter II.) the circumstances in which both Uranus and Neptune were discovered. It should, however, be added that after the discovery of Uranus, that planet was found to have been already noted upon several occasions by different observers, but always without the least suspicion that it was other than a mere faint star. Again, with reference to the discovery of Neptune, it may here be mentioned that the apparent amount by which that planet had pulled Uranus out of its place upon the starry background was exceedingly small--so small, indeed, that no eye could have detected it without the aid of a telescope!

Of the two predictions of the place of Neptune in the sky, that of Le Verrier was the nearer. Indeed, the position calculated by Adams was more than twice as far out. But Adams was by a long way the first in the field with his results, and only for unfortunate delays the prize would certainly have fallen to him. For instance, there was no star-map at Cambridge, and Professor Challis, the director of the observatory there, was in consequence obliged to make a laborious examination of the stars in the suspected region. On the other hand, all that Galle had to do was to compare that part of the sky where Le Verrier told him to look with the Berlin star-chart which he had by him. This he did on September 23, 1846, with the result that he quickly noted an eighth magnitude star which did not figure in that chart. By the next night this star had altered its position in the sky, thus disclosing the fact that it was really a planet.

Six days later Professor Challis succeeded in finding the planet, but of course he was now too late. On reviewing his labours he ascertained that he had actually noted down its place early in August, and had he only been able to sift his observations as he made them, the discovery would have been made then.

Later on it was found that Neptune had only just missed being discovered about fifty years earlier. In certain observations made during 1795, the famous French astronomer, Lalande, found that a star, which he had mapped in a certain position on the 8th of May of that year, was in a different position two days later. The idea of a planet does not appear to have entered his mind, and he merely treated the first observation as an error!

The reader will, no doubt, recollect how the discovery of the asteroids was due in effect to an apparent break in the seemingly regular sequence of the planetary orbits outwards from the sun. This curious sequence of relative distances is usually known as "Bode's Law," because it was first brought into general notice by an astronomer of that name. It had, however, previously been investigated mathematically by Titius in 1772. Long before this, indeed, the unnecessarily wide space between the orbits of Mars and Jupiter had attracted the attention of the great Kepler to such a degree, that he predicted that a planet would some day be found to fill the void. Notwithstanding the service which the so-called Law of Bode has indirectly rendered to astronomy, it has strangely enough been found after all not to rest upon any scientific foundation. It will not account for the distance from the sun of the orbit of Neptune, and the very sequence seems on the whole to be in the nature of a mere coincidence.

Neptune is invisible to the naked eye; Uranus is just at the limit of visibility. Both planets are, however, so far from us that we can get but the poorest knowledge of their condition and surroundings. Uranus, up to the present, is known to be attended by four satellites, and Neptune by one. The planets themselves are about equal in size; their diameters, roughly speaking, being about one-half that of Saturn. Some markings have, indeed, been seen upon the disc of Uranus, but they are very indistinct and fleeting. From observation of them, it is assumed that the planet rotates on its axis in a period of some ten to twelve hours. No definite markings have as yet been seen upon Neptune, which body is described by several observers as resembling a faint planetary nebula.

With regard to their physical condition, the most that can be said about these two planets is that they are probably in much the same vaporous state as Jupiter and Saturn. On account of their great distance from the sun they can receive but little solar heat and light. Seen from Neptune, in fact, the sun would appear only about the size of Venus at her best, though of a brightness sufficiently intense to illumine the Neptunian landscape with about seven hundred times our full moonlight.

[22] Mr. P. Melotte, of Greenwich Observatory, while examining a photograph taken there on February 28, 1908, discovered upon it a very faint object which it is firmly believed will prove to be an eighth satellite of Jupiter. This object was afterwards found on plates exposed as far back as January 27. It has since been photographed several times at Greenwich, and also at Heidelberg (by Dr. Max Wolf) and at the Lick Observatory. Its movement is probably retrograde, like that of Phoebe (p. 240).

[23] In the history of astronomy two salient points stand out.

The first of these is the number of "independent" discoveries which have taken place; such, for instance, as in the cases of Le Verrier and Adams with regard to Neptune, and of Lockyer and Janssen in the matter of the spectroscopic method of observing solar prominences.

The other is the great amount of "anticipation." Copernicus, as we have seen, was anticipated by the Greeks; Kepler was not actually the first who thought of elliptic orbits; others before Newton had imagined an attractive force.

Both these points furnish much food for thought!

CHAPTER XIX

COMETS

The reader has, no doubt, been struck by the marked uniformity which exists among those members of the solar system with which we have dealt up to the present. The sun, the planets, and their satellites are all what we call solid bodies. The planets move around the sun, and the satellites around the planets, in orbits which, though strictly speaking, ellipses, are yet not in any instance of a very oval form. Two results naturally follow from these considerations. Firstly, the bodies in question hide the light coming to us from those further off, when they pass in front of them. Secondly, the planets never get so far from the sun that we lose sight of them altogether.

With the objects known as Comets it is, however, quite the contrary. These objects do not conform to our notions of solidity. They are so transparent that they can pass across the smallest star without dimming its light in the slightest degree. Again, they are only visible to us during a portion of their orbits. A comet may be briefly described as an illuminated filmy-looking object, made up usually of three portions--a head, a nucleus, or brighter central portion within this head, and a tail. The heads of comets vary greatly in size; some, indeed, appear quite small, like stars, while others look even as large as the moon. Occasionally the nucleus is wanting, and sometimes the tail also.

These mysterious visitors to our skies come up into view out of the immensities beyond, move towards the sun at a rapidly increasing speed, and, having gone around it, dash away again into the depths of space. As a comet approaches the sun, its body appears to grow smaller and smaller, while, at the same time, it gradually throws out behind it an appendage like a tail. As the comet moves round the central orb this tail is always directed away from the sun; and when it departs again into space the tail goes in advance. As the comet's distance from the sun increases, the tail gradually shrinks away and the head once more grows in size (see Fig. 18). In consequence of these changes, and of the fact that we lose sight of comets comparatively quickly, one is much inclined to wonder what further changes may take place after the bodies have passed beyond our ken.

The orbits of comets are, as we have seen, very elliptic. In some instances this ellipticity is so great as to take the bodies out into space to nearly six times the distance of Neptune from the sun. For a long time, indeed, it was considered that comets were of two kinds, namely, those which actually belonged to the solar system, and those which were merely visitors to it for the first and only time--rushing in from the depths of space, rapidly circuiting the sun, and finally dashing away into space again, never to return. On the contrary, nowadays, astronomers are generally inclined to regard comets as permanent members of the solar system.

The difficulty, however, of deciding absolutely whether the orbits of comets are really always closed curves, that is to say, curves which must sooner or later bring the bodies back again towards the sun, is, indeed, very great. Comets, in the first place, are always so diffuse, that it is impossible to determine their exact position, or, rather, the exact position of that important point within them, known as the centre of gravity. Secondly, that stretch of its orbit along which we can follow a comet, is such a very small portion of the whole path, that the slightest errors of observation which we make will result in considerably altering our estimate of the actual shape of the orbit.

Comets have been described as so transparent that they can pass across the sky without dimming the lustre of the smallest stars, which the thinnest fog or mist would do. This is, indeed, true of every portion of a comet except the nucleus, which is, as its name implies, the densest part. And yet, in contrast to this ghostlike character, is the strange fact that when comets are of a certain brightness they may actually be seen in full daylight.

As might be gathered from their extreme tenuity, comets are so exceedingly small in mass that they do not appear to exert any gravitational attraction upon the other bodies of our system. It is, indeed, a known fact that in the year 1886 a comet passed right amidst the satellites of Jupiter without disturbing them in the slightest degree. The attraction of the planet, on the other hand, so altered the comet's orbit, as to cause it to revolve around the sun in a period of seven years, instead of twenty-seven, as had previously been the case. Also, in 1779, the comet known as Lexell's passed quite close to Jupiter, and its orbit was so changed by that planet's attraction that it has never been seen since. The density of comets must, as a rule, be very much less than the one-thousandth part of that of the air at the surface of our globe; for, if the density of the comet were even so small as this, its mass would not be inappreciable.

If comets are really undoubted members of the solar system, the circumstances in which they were evolved must have been different from those which produced the planets and satellites. The axial rotations of both the latter, and also their revolutions, take place in one certain direction;[24] their orbits, too, are ellipses which do not differ much from circles, and which, furthermore, are situated fairly in the one plane. Comets, on the other hand, do not necessarily travel round the sun in the same fixed direction as the planets. Their orbits, besides, are exceedingly elliptic; and, far from keeping to one plane, or even near it, they approach the sun from all directions.

Broadly speaking, comets may be divided into two distinct classes, or "families." In the first class, the same orbit appears to be shared in common by a series of comets which travel along it, one following the other. The comets which appeared in the years 1668, 1843, 1880, 1882, and 1887 are instances of a number of different bodies pursuing the same path around the sun. The members of a comet family of this kind are observed to have similar characteristics. The idea is that such comets are merely portions of one much larger cometary body, which became broken up by the gravitational action of other bodies in the system, or through violent encounter with the sun's surroundings.

The second class is composed of comets which are supposed to have been seized by the gravitative action of certain planets, and thus forced to revolve in short ellipses around the sun, well within the limits of the solar system. These comets are, in consequence, spoken of as "captures." They move around the sun in the same direction as the planets do. Jupiter has a fairly large comet family of this kind attached to him. As a result of his overpowering gravitation, it is imagined that during the ages he must have attracted a large number of these bodies on his own account, and, perhaps, have robbed other planets of their captures. His family at present numbers about thirty. Of the other planets, so far as we know, Saturn possesses a comet family of two, Uranus three, and Neptune six. There are, indeed, a few comets which appear as if under the influence of some force situated outside the known bounds of the solar system, a circumstance which goes to strengthen the idea that other planets may revolve beyond the orbit of Neptune. The terrestrial planets, on the other hand, cannot have comet families; because the enormous gravitative action of the sun in their vicinity entirely overpowers the attractive force which they exert upon those comets which pass close to them. Besides this, a comet, when in the inner regions of the solar system, moves with such rapidity, that the gravitational pull of the planets there situated is not powerful enough to deflect it to any extent. It must not be presumed, however, that a comet once captured should always remain a prisoner. Further disturbing causes might unsettle its newly acquired orbit, and send it out again into the celestial spaces.

With regard to the matter of which comets are composed, the spectroscope shows the presence in them of hydrocarbon compounds (a notable characteristic of these bodies), and at times, also, of sodium and iron. Some of the light which we get from comets is, however, merely reflected sunlight.

The fact that the tails of comets are always directed away from the sun, has given rise to the idea that this is caused by some repelling action emanating from the sun itself, which is continually driving off the smallest particles. Two leading theories have been formulated to account for the tails themselves upon the above assumption. One of these, first suggested by Olbers in 1812, and now associated with the name of the Russian astronomer, the late Professor Brikhine, who carefully worked it out, presumes an electrical action emanating from the sun; the other, that of Arrhenius, supposes a pressure exerted by the solar light in its radiation outwards into space. It is possible, indeed, that repelling forces of both these kinds may be at work together. Minute particles are probably being continually produced by friction and collisions among the more solid parts in the heads of comets. Supposing that such particles are driven off altogether, one may therefore assume that the so-called captured comets are disintegrating at a comparatively rapid rate. Kepler long ago maintained that "comets die," and this actually appears to be the case. The ordinary periodic ones, such, for instance, as Encke's Comet, are very faint, and becoming fainter at each return. Certain of these comets have, indeed, failed altogether to reappear. It is notable that the members of Jupiter's comet family are not very conspicuous objects. They have small tails, and even in some cases have none at all. The family, too, does not contain many members, and yet one cannot but suppose that Jupiter, on account of his great mass, has had many opportunities for making captures adown the ages.

Of the two theories to which allusion has above been made, that of Brikhine has been worked out so carefully, and with such a show of plausibility, that it here calls for a detailed description. It appears besides to explain the phenomena of comets' tails so much more satisfactorily than that of Arrhenius, that astronomers are inclined to accept it the more readily of the two. According to Brikhine's theory the electrical repulsive force, which he assumes for the purposes of his argument, will drive the minutest particles of the comet in a direction away from the sun much more readily than the gravitative action of that body will pull them towards it. This may be compared to the ease with which fine dust may be blown upwards, although the earth's gravitation is acting upon it all the time.

The researches of Brikhine, which began seriously with his investigation of Coggia's Comet of 1874, led him to classify the tails of comets in three types. Presuming that the repulsive force emanating from the sun did not vary, he came to the conclusion that the different forms assumed by cometary tails must be ascribed to the special action of this force upon the various elements which happen to be present in the comet. The tails which he classes as of the first type, are those which are long and straight and point directly away from the sun. Examples of such tails are found in the comets of 1811, 1843, and 1861. Tails of this kind, he thinks, are in all probability formed of hydrogen. His second type comprises those which are pointed away from the sun, but at the same time are considerably curved, as was seen in the comets of Donati and Coggia. These tails are formed of hydrocarbon gas. The third type of tail is short, brush-like, and strongly bent, and is formed of the vapour of iron, mixed with that of sodium and other elements. It should, however, be noted that comets have occasionally been seen which possess several tails of these various types.

We will now touch upon a few of the best known comets of modern times.

The comet of 1680 was the first whose orbit was calculated according to the laws of gravitation. This was accomplished by Newton, and he found that the comet in question completed its journey round the sun in a period of about 600 years.

In 1682 there appeared a great comet, which has become famous under the name of Halley's Comet, in consequence of the profound investigations made into its motion by the great astronomer, Edmund Halley. He fixed its period of revolution around the sun at about seventy-five years, and predicted that it would reappear in the early part of 1759. He did not, however, live to see this fulfilled, but the comet duly returned--the first body of the kind to verify such a prediction--and was detected on Christmas Day, 1758, by George Palitzch, an amateur observer living near Dresden. Halley also investigated the past history of the comet, and traced it back to the year 1456. The orbit of Halley's comet passes out slightly beyond the orbit of Neptune. At its last visit in 1835, this comet passed comparatively close to us, namely, within five million miles of the earth. According to the calculations of Messrs P.H. Cowell and A.C.D. Crommelin of Greenwich Observatory, its next return will be in the spring of 1910; the nearest approach to the earth taking place about May 12.

On the 26th of March, 1811, a great comet appeared, which remained visible for nearly a year and a half. It was a magnificent object; the tail being about 100 millions of miles in length, and the head about 127,000 miles in diameter. A detailed study which he gave to this comet prompted Olbers to put forward that theory of electrical repulsion which, as we have seen, has since been so carefully worked out by Brikhine. Olbers had noticed that the particles expelled from the head appeared to travel to the end of the tail in about eleven minutes, thus showing a velocity per second very similar to that of light.

The discovery in 1819 of the comet known as Encke's, because its orbit was determined by an astronomer of that name, drew attention for the first time to Jupiter's comet family, and, indeed, to short-period comets in general. This comet revolves around the sun in the shortest known period of any of these bodies, namely, 3-1/3 years. Encke predicted that it would return in 1822. This duly occurred, the comet passing at its nearest to the sun within three hours of the time indicated; being thus the second instance of the fulfilment of a prediction of the kind. A certain degree of irregularity which Encke's Comet displays in the dates of its returns to the sun, has been supposed to indicate that it passes in the course of its orbit through some retarding medium, but no definite conclusions have so far been arrived at in this matter.

A comet, which appeared in 1826, goes by the name of Biela's Comet, because of its discovery by an Austrian military officer, Wilhelm von Biela. This comet was found to have a period of between six and seven years. Certain calculations made by Olbers showed that, at its return in 1832, it would pass through the earth's orbit. The announcement of this gave rise to a panic; for people did not wait to inquire whether the earth would be anywhere near that part of its orbit when the comet passed. The panic, however, subsided when the French astronomer, Arago, showed that at the moment in question the earth would be some 50 millions of miles away from the point indicated!

From a drawing made on October 9th, 1858, by G.P. Bond, of Harvard College Observatory, U.S.A. A good illustration of Brikhine's theory: note the straight tails of his first type, and the curved tail of his second.

In 1846, shortly after one of its returns, Biela's Comet divided into two portions. At its next appearance (1852) these portions had separated to a distance of about 1-1/2 millions of miles from each other. This comet, or rather its constituents, have never since been seen.

Perhaps the most remarkable comet of recent times was that of 1858, known as Donati's, it having been discovered at Florence by the Italian astronomer, G.B. Donati. This comet, a magnificent object, was visible for more than three months with the naked eye. Its tail was then 54 millions of miles in length. It was found to revolve around the sun in a period of over 2000 years, and to go out in its journey to about 5-1/2 times the distance of

Neptune. Its motion is retrograde, that is to say, in the contrary direction to the usual movement in the solar system. A number of beautiful drawings of Donati's Comet were made by the American astronomer, G.P. Bond. One of the best of these is reproduced on Plate XVII., p. 256.

In 1861 there appeared a great comet. On the 30th of June of that year the earth and moon actually passed through its tail; but no effects were noticed, other than a peculiar luminosity in the sky.

In the year 1881 there appeared another large comet, known as Tebbutt's Comet, from the name of its discoverer. This was the first comet of which a satisfactory photograph was obtained. The photograph in question was taken by the late M. Janssen.

The comet of 1882 was of vast size and brilliance. It approached so close to the sun that it passed through some 100,000 miles of the solar corona. Though its orbit was not found to have been altered by this experience, its nucleus displayed signs of breaking up. Some very fine photographs of this comet were obtained at the Cape of Good Hope by Mr. (now Sir David) Gill.

The comet of 1889 was followed with the telescope nearly up to the orbit of Saturn, which seems to be the greatest distance at which a comet has ever been seen.

The first discovery of a comet by photographic means[25] was made by Professor Barnard in 1892; and, since then, photography has been employed with marked success in the detection of small periodic comets.

The best comet seen in the Northern hemisphere since that of 1882, appears to have been Daniel's Comet of 1907 (see Plate XVIII., p. 258). This comet was discovered on June 9, 1907, by Mr. Z. Daniel, at Princeton Observatory, New Jersey, U.S.A. It became visible to the naked eye about mid-July of that year, and reached its greatest brilliancy about the end of August. It did not, however, attract much popular attention, as its position in the sky allowed it to be seen only just before dawn.

[24] With the exception, of course, of such an anomaly as the retrograde motion of the ninth satellite of Saturn.

[25] If we except the case of the comet which was photographed near the solar corona in the eclipse of 1882.

From a photograph taken, on August 11th, 1907, by Dr. Max Wolf, at the Astrophysical Observatory, Heidelberg. The instrument used was a 28-inch reflecting telescope, and the time of exposure was fifteen minutes. As the telescope was guided to follow the moving comet, the stars have imprinted themselves upon the photographic plate as short trails. This is clearly the opposite to what is depicted on Plate XIII.

(Page 258)]

CHAPTER XX

REMARKABLE COMETS

If eclipses were a cause of terror in past ages, comets appear to have been doubly so. Their much longer continuance in the sight of men had no doubt something to say to this, and also the fact that they arrived without warning; it not being then possible to give even a rough prediction of their return, as in the case of eclipses. As both these phenomena were occasional, and out of the ordinary course of things, they drew exceptional attention as unusual events always do; for it must be allowed that quite as wonderful things exist, but they pass unnoticed merely because men have grown accustomed to them.

For some reason the ancients elected to class comets along with meteors, the aurora borealis, and other phenomena of the atmosphere, rather than with the planets and the bodies of the spaces beyond. The sudden appearance of these objects led them to be regarded as signs sent by the gods to announce remarkable events, chief among these being the deaths of monarchs. Shakespeare has reminded us of this in those celebrated lines in Julius Caesar:--

"When beggars die there are no comets seen, The heavens themselves blaze forth the death of princes."

Numbed by fear, the men of old blindly accepted these presages of fate; and did not too closely question whether the threatened danger was to their own nation or to some other, to their ruler or to his enemy. Now and then, as in the case of the Roman Emperor Vespasian, there was a cynical attempt to apply some reasoning to the portent. That emperor, in alluding to the comet of A.D. 79, is reported to have said: "This hairy star does not concern me; it menaces rather the King of the Parthians, for he is hairy and I am bald." Vespasian, all the same, died shortly afterwards!

Pliny, in his natural history, gives several instances of the terrible significance which the ancients attached to comets. "A comet," he says, "is ordinarily a very fearful star; it announces no small effusion of blood. We have seen an example of this during the civil commotion of Octavius."

A very brilliant comet appeared in 371 B.C., and about the same time an earthquake caused Helic?and Bura, two towns in Achaia, to be swallowed up by the sea. The following remark made by Seneca concerning it shows that the ancients did not consider comets merely as precursors, but even as actual causes of fatal events: "This comet, so anxiously observed by every one, because of the great catastrophe which it produced as soon as it appeared, the submersion of Bura and Helic?"

Comets are by no means rare visitors to our skies, and very few years have elapsed in historical times without such objects making their appearance. In the Dark and Middle Ages, when Europe was split up into many small kingdoms and principalities, it was, of course, hardly possible for a comet to appear without the death of some ruler occurring near the time. Critical situations, too, were continually arising in those disturbed days. The end of Louis le Debonnaire was hastened, as the reader will, no doubt, recollect, by the great eclipse of 840; but it was firmly believed that a comet which had appeared a year or two previously presaged his death. The comet of 1556 is reported to have influenced the abdication of the Emperor Charles V.; but curiously enough, this event had already taken place before the comet made its appearance! Such beliefs, no doubt, had a very real effect upon rulers of a superstitious nature, or in a weak state of health. For instance, Gian Galeazzo Visconti, Duke of Milan, was sick when the comet of 1402 appeared. After seeing it, he is said to have exclaimed: "I render thanks to God for having decreed that my death should be announced to men by this celestial sign."

His malady then became worse, and he died shortly afterwards.

It is indeed not improbable that such superstitious fears in monarchs were fanned by those who would profit by their deaths, and yet did not wish to stain their own hands with blood.

Evil though its effects may have been, this morbid interest which past ages took in comets has proved of the greatest service to our science. Had it not been believed that the appearance of these objects was attended with far-reaching effects, it is very doubtful whether the old chroniclers would have given themselves the trouble of alluding to them at all; and thus the modern investigators of cometary orbits would have lacked a great deal of important material.

We will now mention a few of the most notable comets which historians have recorded.

A comet which appeared in 344 B.C. was thought to betoken the success of the expedition undertaken in that year by Timoleon of Corinth against Sicily. "The gods by an extraordinary prodigy announced his success and future greatness: a burning torch appeared in the heavens throughout the night and preceded the fleet of Timoleon until it arrived off the coast of Sicily."

The comet of 43 B.C. was generally believed to be the soul of Caesar on its way to heaven.

Josephus tells us that in A.D. 69 several prodigies, and amongst them a comet in the shape of a sword, announced the destruction of Jerusalem. This comet is said to have remained over the city for the space of a year!

A comet which appeared in A.D. 336 was considered to have announced the death of the Emperor Constantine.

But perhaps the most celebrated comet of early times was the one which appeared in A.D. 1000. That year was, in more than one way, big with portent, for there had long been a firm belief that the Christian era could not possibly run into four figures. Men, indeed, steadfastly believed that when the thousand years had ended, the millennium would immediately begin.

Therefore they did not reap neither did they sow, they toiled not, neither did they spin, and the appearance of the comet strengthened their convictions. The fateful year, however, passed by without anything remarkable taking place; but the neglect of husbandry brought great famine and pestilence over Europe in the years which followed.

In April 1066, that year fraught with such immense consequences for England, a comet appeared. No one doubted but that it was a presage of the success of the Conquest, and perhaps, indeed, it had its due weight in determining the minds and actions of the men who took part in the expedition. Nova stella, novus rex ("a new star, a new sovereign") was a favourite proverb of the time. The chroniclers, with one accord, have delighted to relate that the Normans, "guided by a comet," invaded England. A representation of this object appears in the Bayeux Tapestry (see Fig. 19, p. 263).[26]

We have mentioned Halley's Comet of 1682, and how it revisits the neighbourhood of the earth at intervals of seventy-six years. The comet of 1066 has for many years been supposed to be Halley's Comet on one of its visits. The identity of these two, however, was only quite recently placed beyond all doubt by the investigations of Messrs Cowell and Crommelin. This comet appeared also in 1456, when John Huniades was defending Belgrade against the Turks led by Mahomet II., the conqueror of Constantinople, and is said to have paralysed both armies with fear.

The Middle Ages have left us descriptions of comets, which show only too well how the imagination will run riot under the stimulus of terror. For instance, the historian, Nicetas, thus describes the comet of the year 1182: "After the Romans were driven from Constantinople a prognostic was seen of the excesses and crimes to which Andronicus was to abandon himself. A comet appeared in the heavens similar to a writhing serpent; sometimes it extended itself, sometimes it drew itself in; sometimes, to the great terror of the spectators, it opened a huge mouth; it seemed that, as if thirsting for human blood, it was upon the point of satiating itself." And, again, the celebrated Ambrose Par? the father of surgery, has left us the following account of the comet of 1528, which appeared in his own time: "This comet," said he, "was so horrible, so frightful, and it produced such great terror in the vulgar, that some died of fear, and others fell sick. It appeared to be of excessive length, and was of the colour of blood. At the summit of it was seen the figure of a bent arm, holding in its hand a great sword, as if about to strike. At the end of the point there were three stars. On both sides of the rays of this comet were seen a great number of axes, knives, blood-coloured swords, among which were a great number of hideous human faces, with beards and bristling hair." Par? it is true, was no astronomer; yet this shows the effect of the phenomenon, even upon a man of great learning, as undoubtedly he was. It should here be mentioned that nothing very remarkable happened at or near the year 1528.

Concerning the comet of 1680, the extraordinary story got about that, at Rome, a hen had laid an egg on which appeared a representation of the comet!

But the superstitions with regard to comets were now nearing their end. The last blow was given by Halley, who definitely proved that they obeyed the laws of gravitation, and circulated around the sun as planets do; and further announced that the comet of 1682 had a period of seventy-six years, which would cause it to reappear in the year 1759. We have seen how this prediction was duly verified. We have seen, too, how this comet appeared again in 1835, and how it is due to return in the early part of 1910.

[26] With regard to the words "Isti mirant stella" in the figure, Mr. W.T. Lynn suggests that they may not, after all, be the grammatically bad Latin which they appear, but that the legend is really "Isti mirantur stellam," the missing letters being supposed to be hidden by the building and the comet.

CHAPTER XXI

METEORS OR SHOOTING STARS

Any one who happens to gaze at the sky for a short time on a clear night is pretty certain to be rewarded with a view of what is popularly known as a "shooting star." Such an object, however, is not a star at all, but has received its appellation from an analogy; for the phenomenon gives to the inexperienced in these matters an impression as if one of the many points of light, which glitter in the vaulted heaven, had suddenly become loosened from its place, and was falling towards the earth. In its passage across the sky the moving object leaves behind a trail of light which usually lasts for a few moments. Shooting stars, or meteors, as they are technically termed, are for the most part very small bodies, perhaps no larger than peas or pebbles, which, dashing towards our earth from space beyond, are heated to a white heat, and reduced to powder by the friction resulting from their rapid passage into our atmosphere. This they enter at various degrees of speed, in some cases so great as 45 miles a second. The speed, of course, will depend greatly upon whether the earth and the meteors are rushing towards each other, or whether the latter are merely overtaking the earth. In the first of these cases the meteors will naturally collide with the atmosphere with great force; in the other case they will plainly come into it with much less rapidity. As has been already stated, it is from observations of such bodies that we are enabled to estimate, though very imperfectly, the height at which the air around our globe practically ceases, and this height is imagined to be somewhere about 100 miles. Fortunate, indeed, is it for us that there is a goodly layer of atmosphere over our heads, for, were this not so, these visitors from space would strike upon the surface of our earth night and day, and render existence still more unendurable than many persons choose to consider it. To what a bombardment must the moon be continually subject, destitute as she is of such an atmospheric shield!

It is only in the moment of their dissolution that we really learn anything about meteors, for these bodies are much too small to be seen before they enter our atmosphere. The d 閣 ris arising from their destruction is wafted over the earth, and, settling down eventually upon its surface, goes to augment the accumulation of that humble domestic commodity which men call dust. This continual addition of material tends, of course, to increase the mass of the earth, though the effect thus produced will be on an exceedingly small scale.

The total number of meteors moving about in space must be practically countless. The number which actually dash into the earth's atmosphere during each year is, indeed, very great. Professor Simon Newcomb, the well-known American astronomer, has estimated that, of the latter, those large enough to be seen with the naked eye cannot be in all less than 146,000,000,000 per annum. Ten times more numerous still are thought to be those insignificant ones which are seen to pass like mere sparks of light across the field of an observer's telescope.

Until comparatively recent times, perhaps up to about a hundred years ago, it was thought that meteors were purely terrestrial phenomena which had their origin in the upper regions of the air. It, however, began to be noticed that at certain periods of the year these moving objects appeared to come from definite areas of the sky. Considerations, therefore, respecting their observed velocities, directions, and altitudes, gave rise to the theory that they are swarms of small bodies travelling around the sun in elongated elliptical orbits, all along the length of which they are scattered, and that the earth, in its annual revolution, rushing through the midst of such swarms at the same epoch each year, naturally entangles many of them in its atmospheric net.

The dates at which the earth is expected to pass through the principal meteor-swarms are now pretty well known. These swarms are distinguished from one another by the direction of the sky from which the meteors seem to arrive. Many of the swarms are so wide that the earth takes days, and even weeks, to pass through them. In some of these swarms, or streams, as they are also called, the meteors are distributed with fair evenness along the entire length of their orbits, so that the earth is greeted with a somewhat similar shower at each yearly encounter. In others, the chief portions are bunched together, so that, in certain years, the display is exceptional (see Fig. 20, p. 269). That part of the heavens from which a shower of meteors is seen to emanate is called the "radiant," or radiant point, because the foreshortened view we get of the streaks of light makes it appear as if they radiated outwards from this point. In observations of these bodies the attention of astronomers is directed to registering the path and speed of each meteor, and to ascertaining the position of the radiant. It is from data such as these that computations concerning the swarms and their orbits are made.

For the present state of knowledge concerning meteors, astronomy is largely indebted to the researches of Mr. W.F. Denning, of Bristol, and of the late Professor A.S. Herschel.

During the course of each year the earth encounters a goodly number of meteor-swarms. Three of these, giving rise to fine displays, are very well known--the "Perseids," or August Meteors, and the "Leonids" and "Bielids," which appear in November.

Of the above three the Leonid display is by far the most important, and the high degree of attention paid to it has laid the foundation of meteoric astronomy in much the same way that the study of the fascinating corona has given such an impetus to our knowledge of the sun. The history of this shower of meteors may be traced back as far as A.D. 902, which was known as the "Year of the Stars." It is related that in that year, on the night of October 12th--the shower now comes about a month later--whilst the Moorish King, Ibrahim Ben Ahmed, lay dying before Cosenza, in Calabria, "a multitude of falling stars scattered themselves across the sky like rain," and the beholders shuddered at what they considered a dread celestial portent. We have, however, little knowledge of the subsequent history of the Leonids until 1698, since which time the maximum shower has appeared with considerable regularity at intervals of about thirty-three years. But it was not until 1799 that they sprang into especial notice. On the 11th November in that year a splendid display was witnessed at Cumana, in South America, by the celebrated travellers, Humboldt and Bonpland. Finer still, and surpassing all displays of the kind ever seen, was that of November 12, 1833, when the meteors fell thick as snowflakes, 240,000 being estimated to have appeared during seven hours. Some of them were even so bright as to be seen in full daylight. The radiant from which the meteors seem to diverge was ascertained to be situated in the head of the constellation of the Lion, or "Sickle of Leo," as it is popularly termed, whence their name--Leonids. It was from a discussion of the observations then made that the American astronomer, Olmsted, concluded that these meteors sprang upon us from interplanetary space, and were not, as had been hitherto thought, born of our atmosphere. Later on, in 1837, Olbers formulated the theory that the bodies in question travelled around the sun in an elliptical orbit, and at the same time he established the periodicity of the maximum shower.

The periodic time of recurrence of this maximum, namely, about thirty-three years, led to eager expectancy as 1866 drew near. Hopes were then fulfilled, and another splendid display took place, of which Sir Robert Ball, who observed it, has given a graphic description in his Story of the Heavens. The display was repeated upon a smaller scale in the two following years. The Leonids were henceforth deemed to hold an anomalous position among meteor swarms. According to theory the earth cut through their orbit at about the same date each year, and so a certain number were then seen to issue from the radiant. But, in addition, after intervals of thirty-three years, as has been seen, an exceptional display always took place; and this state of things was not limited to one year alone, but was repeated at each meeting for about three years running. The further assumption was, therefore, made that the swarm was much denser in one portion of the orbit than elsewhere,[27] and that this congested part was drawn out to such an extent that the earth could pass through the crossing place during several annual meetings, and still find it going by like a long procession (see Fig. 20, p. 269).

In accordance with this ascertained period of thirty-three years, the recurrence of the great Leonid shower was timed to take place on the 15th of November 1899. But there was disappointment then, and the displays which occurred during the few years following were not of much importance. A good deal of comment was made at the time, and theories were accordingly put forward to account for the failure of the great shower. The most probable explanation seems to be, that the attraction of one of the larger planets-- Jupiter perhaps--has diverted the orbit somewhat from its old position, and the earth does not in consequence cut through the swarm in the same manner as it used to do.

The other November display alluded to takes place between the 23rd and 27th of that month. It is called the Andromedid Shower, because the meteors appear to issue from the direction of the constellation of Andromeda, which at that period of the year is well overhead during the early hours of the night. These meteors are also known by the name of Bielids, from a connection which the orbit assigned to them appears to have with that of the well-known comet of Biela.

M. Egenitis, Director of the Observatory of Athens, accords to the Bielids a high antiquity. He traces the shower back to the days of the Emperor Justinian. Theophanes, the Chronicler of that epoch, writing of the famous revolt of Nika in the year A.D. 532, says:--"During the same year a great fall of stars came from the evening till the dawn." M. Egenitis notes another early reference to these meteors in A.D. 752, during the reign of the Eastern Emperor, Constantine Copronymous. Writing of that year, Nicephorus, a Patriarch of Constantinople, has as follows:--"All the stars appeared to be detached from the sky, and to fall upon the earth."

The Bielids, however, do not seem to have attracted particular notice until the nineteenth century. Attention first began to be riveted upon them on account of their suspected connection with Biela's comet. It appeared that the same orbit was shared both by that comet and the Bielid swarm. It will be remembered that the comet in question was not seen after its appearance in 1852. Since that date, however, the Bielid shower has shown an increased activity; which was further noticed to be especially great in those years in which the comet, had it still existed, would be due to pass near the earth.

The third of these great showers to which allusion has above been made, namely, the Perseids, strikes the earth about the 10th of August; for which reason it is known on the Continent under the name of the "tears of St. Lawrence," the day in question being sacred to that Saint. This shower is traceable back many centuries, even as far as the year A.D. 811. The name given to these meteors, "Perseids," arises from the fact that their radiant point is situated in the constellation of Perseus. This shower is, however, not by any means limited to the particular night of August 10th, for meteors belonging to the swarm may be observed to fall in more or less varying quantities from about July 8th to August 22nd. The Perseid meteors sometimes fall at the rate of about sixty per hour. They are noted for their great rapidity of motion, and their trails besides often persist for a minute or two before being disseminated. Unlike the other well-known showers, the radiants of which are stationary, that of the Perseids shifts each night a little in an easterly direction.

The orbit of the Perseids cuts that of the earth almost perpendicularly. The bodies are generally supposed to be the result of the disintegration of an ancient comet which travelled in the same orbit. Tuttle's Comet, which passed close to the earth in 1862, also belongs to this orbit; and its period of revolution is calculated to be 131 years. The Perseids appear to be disseminated all along this great orbit, for we meet them in considerable quantities each year. The bodies in question are in general particularly small. The swarm has, however, like most others, a somewhat denser portion, and through this the earth passed in 1848. The aphelion, or point where the far end of the orbit turns back again towards the sun, is situated right away beyond the path of Neptune, at a distance of forty-eight times that of the earth from the sun. The comet of 1532 also belongs to the Perseid orbit. It revisited the neighbourhood of the earth in 1661, and should have returned in 1789. But we have no record of it in that year; for which omission the then politically disturbed state of Europe may account. If not already disintegrated, this comet is due to return in 1919.

This supposed connection between comets and meteor-swarms must be also extended to the case of the Leonids. These meteors appear to travel along the same track as Tempel's Comet of 1866.

It is considered that the attractions of the various bodies of the solar system upon a meteor swarm must eventually result in breaking up the "bunched" portion, so that in time the individual meteors should become distributed along the whole length of the orbit. Upon this assumption the Perseid swarm, in which the meteors are fairly well scattered along its path, should be of greater age than the Leonid. As to the Leonid swarm itself, Le Verrier held that it was first brought into the solar system in A.D. 126, having been captured from outer space by the gravitative action of the planet Uranus.

The acknowledged theory of meteor swarms has naturally given rise to an idea, that the sunlight shining upon such a large collection of particles ought to render a swarm visible before its collision with the earth's atmosphere. Several attempts have therefore been made to search for approaching swarms by photography, but, so far, it appears without success. It has also been proposed, by Mr. W.H.S. Monck, that the stars in those regions from which swarms are due, should be carefully watched, to see if their light exhibits such temporary diminutions as would be likely to arise from the momentary interposition of a cloud of moving particles.

Between ten and fifteen years ago it happened that several well-known observers, employed in telescopic examination of the sun and moon, reported that from time to time they had seen small dark bodies, sometimes singly, sometimes in numbers, in passage across the discs of the luminaries. It was concluded that these were meteors moving in space beyond the atmosphere of the earth. The bodies were called "dark meteors," to emphasise the fact that they were seen in their natural condition, and not in that momentary one in which they had hitherto been always seen; i.e. when heated to white heat, and rapidly vaporised, in the course of their passage through the upper regions of our air. This "discovery" gave promise of such assistance to meteor theories, that calculations were made from the directions in which they had been seen to travel, and the speeds at which they had moved, in the hope that some information concerning their orbits might be revealed. But after a while some doubt began to be thrown upon their being really meteors, and eventually an Australian observer solved the mystery. He found that they were merely tiny particles of dust, or of the black coating on the inner part of the tube of the telescope, becoming detached from the sides of the eye-piece and falling across the field of view. He was led to this conclusion by having noted that a gentle tapping of his instrument produced the "dark" bodies in great numbers! Thus the opportunity of observing meteors beyond our atmosphere had once more failed.

Meteorites, also known as fireballs, are usually placed in quite a separate category from meteors. They greatly exceed the latter in size, are comparatively rare, and do not appear in any way connected with the various showers of meteors. The friction of their passage through the atmosphere causes them to shine with a great light; and if not shattered to pieces by internal explosions, they reach the ground to bury themselves deep in it with a great rushing and noise. When found by uncivilised peoples, or savages, they are, on account of their celestial origin, usually regarded as objects of wonder and of worship, and thus have arisen many mythological legends and deifications of blackened stones. On the other hand, when they get into the possession of the civilised, they are subjected to careful examinations and tests in chemical laboratories. The bodies are, as a rule, composed of stone, in conjunction with iron, nickel, and such elements as exist in abundance upon our earth; though occasionally specimens are found which are practically pure metal. In the museums of the great capitals of both Continents are to be seen some fine collections of meteorites. Several countries--Greenland and Mexico, for instance--contain in the soil much meteoric iron, often in masses so large as to baffle all attempts at removal. Blocks of this kind have been known to furnish the natives in their vicinity for many years with sources of workable iron.

The largest meteorite in the world is one known as the Anighito meteorite. It was brought to the United States by the explorer Peary, who found it at Cape York in Greenland. He estimates its weight at from 90 to 100 tons. One found in Mexico, called the Bacubirito, comes next, with an estimated weight of 27-1/2 tons. The third in size is the Willamette meteorite, found at Willamette in Oregon in 1902. It measures 10 ?6-1/2 ?4-1/2 feet, and weighs about 15-1/2 tons.

[27] The "gem" of the meteor ring, as it has been termed.

CHAPTER XXII

THE STARS

In the foregoing chapters we have dealt at length with those celestial bodies whose nearness to us brings them into our especial notice. The entire room, however, taken up by these bodies, is as a mere point in the immensities of star-filled space. The sun, too, is but an ordinary star; perhaps quite an insignificant one[28] in comparison with the majority of those which stud that background of sky against which the planets are seen to perform their wandering courses.

Dropping our earth and the solar system behind, let us go afield and explore the depths of space.

We have seen how, in very early times, men portioned out the great mass of the so-called "fixed stars" into divisions known as constellations. The various arrangements, into which the brilliant points of light fell as a result of perspective, were noticed and roughly compared with such forms as were familiar to men upon the earth. Imagination quickly saw in them the semblances of heroes and of mighty fabled beasts; and, around these monstrous shapes, legends were woven, which told how the great deeds done in the misty dawn of historical time had been enshrined by the gods in the sky as an example and a memorial for men. Though the centuries have long outlived such fantasies, yet the constellation figures and their ancient names have been retained to this day, pretty well unaltered for want of any better arrangement. The Great and Little Bears, Cassiopeia, Perseus, and Andromeda, Orion and the rest, glitter in our night skies just as they did centuries and centuries ago.

Many persons seem to despair of gaining any real knowledge of astronomy, merely because they are not versed in recognising the constellations. For instance, they will say:--"What is the use of my reading anything about the subject? Why, I believe I couldn't even point out the Great Bear, were I asked to do so!" But if such persons will only consider for a moment that what we call the Great Bear has no existence in fact, they need not be at all disheartened. Could we but view this familiar constellation from a different position in space, we should perhaps be quite unable to recognise it. Mountain masses, for instance, when seen from new directions, are often unrecognisable.

It took, as we have seen, a very long time for men to acknowledge the immense distances of the stars from our earth. Their seeming unchangeableness of position was, as we have seen, largely responsible for the idea that the earth was immovable in space. It is a wonder that the Copernican system ever gained the day in the face of this apparent fixity of the stars. As time went on, it became indeed necessary to accord to these objects an almost inconceivable distance, in order to account for the fact that they remained apparently quite undisplaced, notwithstanding the journey of millions of miles which the earth was now acknowledged to make each year around the sun. In the face of the gradual and immense improvement in telescopes, this apparent immobility of the stars was, however, not destined to last. The first ascertained displacement of a star, namely that of 61 Cygni, noted by Bessel in the year 1838, definitely proved to men the truth of the Copernican system. Since then some forty more stars have been found to show similar tiny displacements. We are, therefore, in possession of the fact, that the actual distances of a few out of the great host can be calculated.

To mention some of these. The nearest star to the earth, so far as we yet know, is Alpha Centauri, which is distant from us about 25 billions of miles. The light from this star, travelling at the stupendous rate of about 186,000 miles per second, takes about 4-1/4 years to reach our earth, or, to speak astronomically, Alpha Centauri is about 4-1/4 "light years" distant from us. Sirius--the brightest star in the whole sky--is at twice this distance, i.e. about 8-1/2 light years. Vega is about 30 light years distant from us, Capella about 32, and Arcturus about 100.

The displacements, consequent on the earth's movement, have, however, plainly nothing to say to any real movements on the part of the stars themselves. The old idea was that the stars were absolutely fixed; hence arose the term "fixed stars"--a term which, though inaccurate, has not yet been entirely banished from the astronomical vocabulary. But careful observations extending over a number of years have shown slight changes of position among these bodies; and such alterations cannot be ascribed to the revolution of the earth in its orbit, for they appear to take place in every direction. These evidences of movement are known as "proper motions," that is to say, actual motions in space proper to the stars themselves. Stars which are comparatively near to us show, as a rule, greater proper motions than those which are farther off. It must not, however, be concluded that these proper motions are of any very noticeable amounts. They are, as a matter of fact, merely upon the same apparently minute scale as other changes in the heavens; and would largely remain unnoticed were it not for the great precision of modern astronomical instruments.

One of the swiftest moving of the stars is a star of the sixth magnitude in the constellation of the Great Bear; which is known as "1830 Groombridge," because this was the number assigned to it in a catalogue of stars made by an astronomer of that name. It is popularly known as the "Runaway Star," a name given to it by Professor Newcomb. Its speed is estimated to be at least 138 miles per second. It may be actually moving at a much greater rate, for it is possible that we see its path somewhat foreshortened.

A still greater proper motion--the greatest, in fact, known--is that of an eighth magnitude star in the southern hemisphere, in the constellation of Pictor. Nothing, indeed, better shows the enormous distance of the stars from us, and the consequent inability of even such rapid movements to alter the appearance of the sky during the course of ages, than the fact that it would take more than two centuries for the star in question to change its position in the sky by a space equal to the apparent diameter of the moon; a statement which is equivalent to saying that, were it possible to see this star with the naked eye, which it is not, at least twenty-five years would have to elapse before one would notice that it had changed its place at all!

Both the stars just mentioned are very faint. That in Pictor is, as has been said, not visible to the naked eye. It appears besides to be a very small body, for Sir David Gill finds a parallax which makes it only as far off from us as Sirius. The Groombridge star, too, is just about the limit of ordinary visibility. It is, indeed, a curious fact that the fainter stars seem, on the average, to be moving more rapidly than the brighter.

Investigations into proper motions lead us to think that every one of the stars must be moving in space in some particular direction. To take a few of the best known. Sirius and Vega are both approaching our system at a rate of about 10 miles per second, Arcturus at about 5 miles per second, while Capella is receding from us at about 15 miles per second. Of the twin brethren, Castor and Pollux, Castor is moving away from us at about 4-1/2 miles per second, while Pollux is coming towards us at about 33 miles per second.

Much of our knowledge of proper motions has been obtained indirectly by means of the spectroscope, on the Doppler principle already treated of, by which we are enabled to ascertain whether a source from which light is coming is approaching or receding.

The sun being, after all, a mere star, it will appear only natural for it also to have a proper motion of its own. This is indeed the case; and it is rushing along in space at a rate of between ten and twelve miles per second, carrying with it its whole family of planets and satellites, of comets and meteors. The direction in which it is advancing is towards a point in the constellation of Lyra, not far from its chief star Vega. This is shown by the fact that the stars about the region in question appear to be opening out slightly, while those in the contrary portion of the sky appear similarly to be closing together.

Sir William Herschel was the first to discover this motion of the sun through space; though in the idea that such a movement might take place he seems to have been anticipated by Mayer in 1760, by Michell in 1767, and by Lalande in 1776.

A suggestion has been made that our solar system, in its motion through the celestial spaces, may occasionally pass through regions where abnormal magnetic conditions prevail, in consequence of which disturbances may manifest themselves throughout the system at the same instant. Thus the sun may be getting the credit of producing what it merely reacts to in common with the rest of its family. But this suggestion, plausible though it may seem, will not explain why the magnetic disturbances experienced upon our earth show a certain dependence upon such purely local facts, as the period of the sun's rotation, for instance.

One would very much like to know whether the movement of the sun is along a straight line, or in an enormous orbit around some centre. The idea has been put forward that it may be moving around the centre of gravity of the whole visible stellar universe. M 鎗 ler, indeed, propounded the notion that Alcyone--the chief star in the group known as the Pleiades--occupied this centre, and that everything revolved around it. He went even further to proclaim that here was the Place of the Almighty, the Mansion of the Eternal! But M 鎗 ler's ideas upon this point have long been shelved.

To return to the general question of the proper motion of stars.

In several instances these motions appear to take place in groups, as if certain stars were in some way associated together. For example, a large number of the stars composing the Pleiades appear to be moving through space in the same direction. Also, of the seven stars composing the Plough, all but two--the star at the end of its "handle," and that one of the "pointers," as they are called, which is the nearer to the pole star--have a common proper motion, i.e. are moving in the same direction and nearly at the same rate.

Further still, the well-known Dutch astronomer, Professor Kapteyn, of Groningen, has lately reached the astonishing conclusion that a great part of the visible universe is occupied by two vast streams of stars travelling in opposite directions. In both these great streams, the individual bodies are found, besides, to be alike in design, alike in chemical constitution, and alike in the stage of their development.

A fable related by the Persian astronomer, Al Sufi (tenth century, A.D.) shows well the changes in the face of the sky which proper motions are bound to produce after great lapses of time. According to this fable the stars Sirius and Procyon were the sisters of the star Canopus. Canopus married Rigel (another star,) but, having murdered her, he fled towards the South Pole, fearing the anger of his sisters. The fable goes on to relate, among other things, that Sirius followed him across the Milky Way. Mr. J. E. Gore, in commenting on the story, thinks that it may be based upon a tradition of Sirius having been seen by the men of the Stone Age on the opposite side of the Milky Way to that on which it now is.

Sirius is in that portion of the heavens from which the sun is advancing. Its proper motion is such that it is gaining upon the earth at the rate of about ten miles per second, and so it must overtake the sun after the lapse of great ages. Vega, on the other hand, is coming towards us from that part of the sky towards which the sun is travelling. It should be about half a million years before the sun and Vega pass by one another. Those who have specially investigated this question say that, as regards the probability of a near approach, it is much more likely that Vega will be then so far to one side of the sun, that her brightness will not be much greater than it is at this moment.

Considerations like these call up the chances of stellar collisions. Such possibilities need not, however, give rise to alarm; for the stars, as a rule, are at such great distances from each other, that the probability of relatively near approaches is slight.

We thus see that the constellations do not in effect exist, and that there is in truth no real background to the sky. We find further that the stars are strewn through space at immense distances from each other, and are moving in various directions hither and thither. The sun, which is merely one of them, is moving also in a certain direction, carrying the solar system along with it. It seems, therefore, but natural to suppose that many a star may be surrounded by some planetary system in a way similar to ours, which accompanies it through space in the course of its celestial journeyings.

[28] Vega, for instance, shines one hundred times more brightly than the sun would do, were it to be removed to the distance at which that star is from us.

CHAPTER XXIII

THE STARS--continued

The stars appear to us to be scattered about the sky without any orderly arrangement. Further, they are of varying degrees of brightness; some being extremely brilliant, whilst others can but barely be seen. The brightness of a star may arise from either of two causes. On the one hand, the body may be really very bright in itself; on the other hand, it may be situated comparatively near to us. Sometimes, indeed, both these circumstances may come into play together.

Since variation in brightness is the most noticeable characteristic of the stars, men have agreed to class them in divisions called "magnitudes." This term, it must be distinctly understood, is employed in such classification without any reference whatever to actual size, being merely taken to designate roughly the amount of light which we receive from a star. The twenty brightest stars in the sky are usually classed in the first magnitude. In descending the scale, each magnitude will be noticed to contain, broadly speaking, three times as many stars as the one immediately above it. Thus the second magnitude contains 65, the third 190, the fourth 425, the fifth 1100, and the sixth 3200. The last of these magnitudes is about the limit of the stars which we are able to see with the naked eye. Adding, therefore, the above numbers together, we find that, without the aid of the telescope, we cannot see more than about 5000 stars in the entire sky--northern and southern hemispheres included. Quite a small telescope will, however, allow us to see down to the ninth magnitude, so that the total number of stars visible to us with such very moderate instrumental means will be well over 100,000.

It must not, however, be supposed that the stars included within each magnitude are all of exactly the same brightness. In fact, it would be difficult to say if there exist in the whole sky two stars which send us precisely the same amount of light. In arranging the magnitudes, all that was done was to make certain broad divisions, and to class within them such stars as were much on a par with regard to brightness. It may here be noted that a standard star of the first magnitude gives us about one hundred times as much light as a star of the sixth magnitude, and about one million times as much as one of the sixteenth magnitude--which is near the limit of what we can see with the very best telescope.

Though the first twenty stars in the sky are popularly considered as being of the first magnitude, yet several of them are much brighter than an average first magnitude star would be. For instance, Sirius--the brightest star in the whole sky--is equal to about eleven first magnitude stars, like, say, Aldebaran. In consequence of such differences, astronomers are agreed in classifying the brightest of them as brighter than the standard first magnitude star. On this principle Sirius would be about two and a half magnitudes above the first. This notation is usefully employed in making comparisons between the amount of light which we receive from the sun, and that which we get from an individual star. Thus the sun will be about twenty-seven and a half magnitudes above the first magnitude. The range, therefore, between the light which we receive from the sun (considered merely as a very bright star) and the first magnitude stars is very much greater than that between the latter and the faintest star which can be seen with the telescope, or even registered upon the photographic plate.

To classify stars merely by their magnitudes, without some definite note of their relative position in the sky, would be indeed of little avail. We must have some simple method of locating them in the memory, and the constellations of the ancients here happily come to our aid. A system combining magnitudes with constellations was introduced by Bayer in 1603, and is still adhered to. According to this the stars in each constellation, beginning with the brightest star, are designated by the letters of the Greek alphabet taken in their usual order. For example, in the constellation of Canis Major, or the Greater Dog, the brightest star is the well-known Sirius, called by the ancients the "Dog Star"; and this star, in accordance with Bayer's method, has received the Greek letter [a] (alpha), and is consequently known as Alpha Canis Majoris.[29] As soon as the Greek letters are used up in this way the Roman alphabet is brought into requisition, after which recourse is had to ordinary numbers.

Notwithstanding this convenient arrangement, some of the brightest stars are nearly always referred to by certain proper names given to them in old times. For instance, it is more usual to speak of Sirius, Arcturus, Vega, Capella, Procyon, Aldebaran, Regulus, and so on, than of [a] Canis Majoris, [a] Bois, [a] Lyr? [a] Aurig? [a] Canis Minoris, [a] Tauri, [a] Leonis, &c. &c.

In order that future generations might be able to ascertain what changes were taking place in the face of the sky, astronomers have from time to time drawn up catalogues of stars. These lists have included stars of a certain degree of brightness, their positions in the sky being noted with the utmost accuracy possible at the period. The earliest known catalogue of this kind was made, as we have seen, by the celebrated Greek astronomer, Hipparchus, about the year 125 B.C. It contained 1080 stars. It was revised and brought up to date by Ptolemy in A.D. 150. Another celebrated list was that drawn up by the Persian astronomer, Al Sufi, about the year A.D. 964. In it 1022 stars were noted down. A catalogue of 1005 stars was made in 1580 by the famous Danish astronomer, Tycho Brahe. Among modern catalogues that of

Argelander (1799-1875) contained as many as 324,198 stars. It was extended by Schofeld so as to include a portion of the Southern Hemisphere, in which way 133,659 more stars were added.

In recent years a project was placed on foot of making a photographic survey of the sky, the work to be portioned out among various nations. A great part of this work has already been brought to a conclusion. About 15,000,000 stars will appear upon the plates; but, so far, it has been proposed to catalogue only about a million and a quarter of the brightest of them. This idea of surveying the face of the sky by photography sprang indirectly from the fine photographs which Sir David Gill took, when at the Cape of Good Hope, of the Comet of 1882. The immense number of star-images which had appeared upon his plates suggested the idea that photography could be very usefully employed to register the relative positions of the stars.

The arrangement of seven stars known as the "Plough" is perhaps the most familiar configuration in the sky (see Plate XIX., p. 292). In the United States it is called the "Dipper," on account of its likeness to the outline of a saucepan, or ladle. "Charles' Wain" was the old English name for it, and readers of Caesar will recollect it under Septentriones, or the "Seven Stars," a term which that writer uses as a synonym for the North. Though identified in most persons' minds with Ursa Major, or the Great Bear, the Plough is actually only a small portion of that famous constellation. Six out of the seven stars which go to make up the well-known figure are of the second magnitude, while the remaining one, which is the middle star of the group, is of the third.

The Greek letters, as borne by the individual stars of the Plough, are a plain transgression of Bayer's method as above described, for they have certainly not been allotted here in accordance with the proper order of brightness. For instance, the third magnitude star, just alluded to as being in the middle of the group, has been marked with the Greek letter [d] (Delta); and so is made to take rank before the stars composing what is called the "handle" of the Plough, which are all of the second magnitude. Sir William Herschel long ago drew attention to the irregular manner in which Bayer's system had been applied. It is, indeed, a great pity that this notation was not originally worked out with greater care and correctness; for, were it only reliable, it would afford great assistance to astronomers in judging of what changes in relative brightness have taken place among the stars.

Though we may speak of using the constellations as a method of finding our way about the sky, it is, however, to certain marked groupings in them of the brighter stars that we look for our sign-posts.

Most of the constellations contain a group or so of noticeable stars, whose accidental arrangement dimly recalls the outline of some familiar geometrical figure and thus arrests the attention.[30] For instance, in an almost exact line with the two front stars of the Plough, or "pointers" as they are called,[31] and at a distance about five times as far away as the interval between them, there will be found a third star of the second magnitude. This is known as Polaris, or the Pole Star, for it very nearly occupies that point of the heaven towards which the north pole of the earth's axis is at present directed (see Plate XIX., p. 292). Thus during the apparently daily rotation of the heavens, this star looks always practically stationary. It will, no doubt, be remembered how Shakespeare has put into the mouth of Julius Caesar these memorable words:--

"But I am constant as the northern star, Of whose true-fix'd and resting quality There is no fellow in the firmament."

We see here the Plough, the Pole Star, Ursa Minor, Auriga, Cassiopeia's Chair, and Lyra. Also the Circle of Precession, along which the Pole makes a complete revolution in a period of 25,868 years, and the Temporary Star discovered by Tycho Brahe in the year 1572.

(Page 291)]

On account of the curvature of the earth's surface, the height at which the Pole Star is seen above the horizon at any place depends regularly upon the latitude; that is to say, the distance of the place in question from the equator. For instance, at the north pole of the earth, where the latitude is greatest, namely, 90? the Pole Star will appear directly overhead; whereas in England, where the latitude is about 50? it will be seen a little more than half way up the northern sky. At the equator, where the latitude is nil, the Pole Star will be on the horizon due north.

In consequence of its unique position, the Pole Star is of very great service in the study of the constellations. It is a kind of centre around which to hang our celestial ideas--a starting point, so to speak, in our voyages about the sky.

According to the constellation figures, the Pole Star is in Ursa Minor, or the Little Bear, and is situated at the end of the tail of that imaginary figure (see Plate XIX., p. 292). The chief stars of this constellation form a group not unlike the Plough, except that the "handle" is turned in the contrary direction. The Americans, in consequence, speak of it as the "Little Dipper."

Before leaving this region of the sky, it will be well to draw attention to the second magnitude star [z] in the Great Bear (Zeta Urs?Majoris), which is the middle star in the "handle" of the Plough. This star is usually known as Mizar, a name given to it by the Arabians. A person with good eyesight can see quite near to it a fifth magnitude star, known under the name of Alcor. We have here a very good example of that deception in the estimation of objects in the sky, which has been alluded to in an earlier chapter. Alcor is indeed distant from Mizar by about one-third the apparent diameter of the moon, yet no one would think so!

On the other side of Polaris from the Plough, and at about an equal apparent distance, will be found a figure in the form of an irregular "W", made up of second and third magnitude stars. This is the well-known "Cassiopeia's Chair"--portion of the constellation of Cassiopeia (see Plate XIX., p. 292).

On either side of the Pole Star, about midway between the Plough and Cassiopeia's Chair, but a little further off from it than these, are the constellations of Auriga and Lyra (see Plate XIX., p. 292). The former constellation will be easily recognised, because its chief features are a brilliant yellowish first magnitude star, with one of the second magnitude not far from it. The first magnitude star is Capella, the other is [b] Aurig? Lyra contains only one first magnitude star--Vega, pale blue in colour. This star has a certain interest for us from the fact that, as a consequence of that slow shift of direction of the earth's axis known as Precession, it will be very near the north pole of the heavens in some 12,000 years, and so will then be considered the pole star (see Plate XIX., p. 292). The constellation of Lyra itself, it must also be borne in mind, occupies that region of the heavens towards which the solar system is travelling.

The handle of the Plough points roughly towards the constellation of Boes, in which is the brilliant first magnitude star Arcturus. This star is of an orange tint.

Between Boes and Lyra lie the constellations of Corona Borealis (or the Northern Crown) and Hercules. The chief feature of Corona Borealis, which is a small constellation, is a semicircle of six small stars, the brightest of which is of the second magnitude. The constellation of Hercules is very extensive, but contains no star brighter than the third magnitude.

Near to Lyra, on the side away from Hercules, are the constellations of Cygnus and Aquila. Of the two, the former is the nearer to the Pole Star, and will be recognised by an arrangement of stars widely set in the form of a cross, or perhaps indeed more like the framework of a boy's kite. The position of Aquila will be found through the fact that three of its brightest stars are almost in a line and close together. The middle of these is Altair, a yellowish star of the first magnitude.

At a little distance from Ursa Major, on the side away from the Pole Star, is the constellation of Leo, or the Lion. Its chief feature is a series of seven stars, supposed to form the head of that animal. The arrangement of these stars is, however, much more like a sickle, wherefore this portion of the constellation is usually known as the "Sickle of Leo." At the end of the handle of the sickle is a white first magnitude star--Regulus.

The reader will, no doubt, recollect that it is from a point in the Sickle of Leo that the Leonid meteors appear to radiate.

The star second in brightness in the constellation of Leo is known as Denebola. This star, now below the second magnitude, seems to have been very much brighter in the past. It is noted, indeed, as a brilliant first magnitude star by Al Sufi, that famous Persian astronomer who lived, as we have seen, in the tenth century. Ptolemy also notes it as of the first magnitude.

In the neighbourhood of Auriga, and further than it from the Pole Star, are several remarkable constellations--Taurus, Orion, Gemini, Canis Minor, and Canis Major (see Plate XX., p. 296).

The first of these, Taurus (or the Bull), contains two conspicuous star groups--the Pleiades and the Hyades. The Pleiades are six or seven small stars quite close together, the majority of which are of the fourth magnitude. This group is sometimes occulted by the moon. The way in which the stars composing it are arranged is somewhat similar to that in the Plough, though of course on a scale ever so much smaller. The impression which the group itself gives to the casual glance is thus admirably pictured in Tennyson's Locksley Hall:--

"Many a night I saw the Pleiads, rising through the mellow shade, Glitter like a swarm of fire-flies tangled in a silver braid."

We see here that magnificent region of the sky which contains the brightest star of all--Sirius. Note also especially the Milky Way, the Pleiades, the Hyades, and the "Belt" and "Sword" of Orion.

The group of the Hyades occupies the "head" of the Bull, and is much more spread out than that of the Pleiades. It is composed besides of brighter stars, the brightest being one of the first magnitude, Aldebaran. This star is of a red colour, and is sometimes known as the "Eye of the Bull."

The constellation of Orion is easily recognised as an irregular quadrilateral formed of four bright stars, two of which, Betelgeux (reddish) and Rigel (brilliant white), are of the first magnitude. In the middle of the quadrilateral is a row of three second magnitude stars, known as the "Belt" of Orion. Jutting off from this is another row of stars called the "Sword" of Orion.

The constellation of Gemini, or the Twins, contains two bright stars--Castor and Pollux--close to each other. Pollux, though marked with the Greek letter [b], is the brighter of the two, and nearly of the standard first magnitude.

Just further from the Pole than Gemini, is the constellation of Canis Minor, or the Lesser Dog. Its chief star is a white first magnitude one--Procyon.

Still further again from the Pole than Canis Minor is the constellation of

Canis Major, or the Greater Dog. It contains the brightest star in the whole sky, the first magnitude star Sirius, bluish-white in colour, also known as the "Dog Star." This star is almost in line with the stars forming the Belt of Orion, and is not far from that constellation.

Taken in the following order, the stars Capella, [b] Aurig? Castor, Pollux, Procyon, and Sirius, when they are all above the horizon at the same time, form a beautiful curve stretching across the heaven.

The groups of stars visible in the southern skies have by no means the same fascination for us as those in the northern. The ancients were in general unacquainted with the regions beyond the equator, and so their scheme of constellations did not include the sky around the South Pole of the heavens. In modern times, however, this part of the celestial expanse was also portioned out into constellations for the purpose of easy reference; but these groupings plainly lack that simplicity of conception and legendary interest which are so characteristic of the older ones.

The brightest star in the southern skies is found in the constellation of Argo, and is known as Canopus. In brightness it comes next to Sirius, and so is second in that respect in the entire heaven. It does not, however, rise above the English horizon.

Of the other southern constellations, two call for especial notice, and these adjoin each other. One is Centaurus (or the Centaur), which contains the two first magnitude stars, [a] and [b] Centauri. The first of these, Alpha Centauri, comes next in brightness to Canopus, and is notable as being the nearest of all the stars to our earth. The other constellation is called Crux, and contains five stars set in the form of a rough cross, known as the "Southern Cross." The brightest of these, [a] Crucis, is of the first magnitude.

Owing to the Precession of the Equinoxes, which, as we have seen, gradually shifts the position of the Pole among the stars, certain constellations used to be visible in ancient times in more northerly latitudes than at present. For instance, some five thousand years ago the Southern Cross rose above the English horizon, and was just visible in the latitude of London. It has, however, long ago even ceased to be seen in the South of Europe. The constellation of Crux happens to be situated in that remarkable region of the southern skies, in which are found the stars Canopus and Alpha Centauri, and also the most brilliant portion of the Milky Way. It is believed to be to this grand celestial region that allusion is made in the Book of Job (ix. 9), under the title of the "Chambers of the South." The "Cross" must have been still a notable feature in the sky of Palestine in the days when that ancient poem was written.

There is no star near enough to the southern pole of the heavens to earn the distinction of South Polar Star.

The Galaxy, or Milky Way (see Plate XX., p. 296), is a broad band of diffused light which is seen to stretch right around the sky. The telescope, however, shows it to be actually composed of a great host of very faint stars--too faint, indeed, to be separately distinguished with the naked eye. Along a goodly stretch of its length it is cleft in two; while near the south pole of the heavens it is entirely cut across by a dark streak.

In this rapid survey of the face of the sky, we have not been able to do more than touch in the broadest manner upon some of the most noticeable star groups and a few of the most remarkable stars. To go any further is not a part of our purpose; our object being to deal with celestial bodies as they actually are, and not in those groupings under which they display themselves to us as a mere result of perspective.

[29] Attention must here be drawn to the fact that the name of the constellation is always put in the genitive case.

[30] The early peoples, as we have seen, appear to have been attracted by those groupings of the stars which reminded them in a way of the figures of men and animals. We moderns, on the other hand, seek almost instinctively for geometrical arrangements. This is, perhaps, symptomatic of the evolution of the race. In the growth of the individual we find, for example, something analogous. A child, who has been given pencil and paper, is almost certain to produce grotesque drawings of men and animals; whereas the idle and half-conscious scribblings which a man may make upon his blotting-paper are usually of a geometrical character.

[31] Because the line joining them points in the direction of the Pole Star.

CHAPTER XXIV

SYSTEMS OF STARS

Many stars are seen comparatively close together. This may plainly arise from two reasons. Firstly, the stars may happen to be almost in the same line of sight; that is to say, seen in nearly the same direction; and though one star may be ever so much nearer to us than the other, the result will give all the appearance of a related pair. A seeming arrangement of two stars in this way is known as a "double," or double star; or, indeed, to be very precise, an "optical double." Secondly, in a pair of stars, both bodies may be about the same distance from us, and actually connected as a system like, for instance, the moon and the earth. A pairing of stars in this way, though often casually alluded to as a double star, is properly termed a "binary," or binary system.

But collocations of stars are by no means limited to two. We find, indeed, all over the sky such arrangements in which there are three or more stars; and these are technically known as "triple" or "multiple" stars respectively. Further, groups are found in which a great number of stars are closely massed together, such a massing together of stars being known as a "cluster."

The Pole Star (Polaris) is a double star, one of the components being of a little below the second magnitude, and the other a little below the ninth. They are so close together that they appear as one star to the naked eye, but they may be seen separate with a moderately sized telescope. The brighter star is yellowish, and the faint one white. This brighter star is found by means of the spectroscope to be actually composed of three stars so very close together that they cannot be seen separately even with a telescope. It is thus a triple star, and the three bodies of which it is composed are in circulation about each other. Two of them are darker than the third.

The method of detecting binary stars by means of the spectroscope is an application of Doppler's principle. It will, no doubt, be remembered that, according to the principle in question, we are enabled, from certain shiftings of the lines in the spectrum of a luminous body, to ascertain whether that body is approaching us or receding from us. Now there are certain stars which always appear single even in the largest telescopes, but when the spectroscope is directed to them a spectrum with two sets of lines is seen.

Such stars must, therefore, be double. Further, if the shiftings of the lines, in a spectrum like this, tell us that the component stars are making small movements to and from us which go on continuously, we are therefore justified in concluding that these are the orbital revolutions of a binary system greatly compressed by distance. Such connected pairs of stars, since they cannot be seen separately by means of any telescope, no matter how large, are known as "spectroscopic binaries."

In observations of spectroscopic binaries we do not always get a double spectrum. Indeed, if one of the components be below a certain magnitude, its spectrum will not appear at all; and so we are left in the strange uncertainty as to whether this component is merely faint or actually dark. It is, however, from the shiftings of the lines in the spectrum of the other component that we see that an orbital movement is going on, and are thus enabled to conclude that two bodies are here connected into a system, although one of these bodies resolutely refuses directly to reveal itself even to the all-conquering spectroscope.

Mizar, that star in the handle of the Plough to which we have already drawn attention, will be found with a small telescope to be a fine double, one of the components being white and the other greenish. Actually, however, as the American astronomer, Professor F.R. Moulton, points out, these stars are so far from each other that if we could be transferred to one of them we should see the other merely as an ordinary bright star. The spectroscope shows that the brighter of these stars is again a binary system of two huge suns, the components revolving around each other in a period of about twenty days. This discovery made by Professor E.C. Pickering, the first of the kind by means of the spectroscope, was announced in 1889 from the Harvard Observatory in the United States.

A star close to Vega, known as [e] (Epsilon) Lyr?(see Plate XIX., p. 292), is a double, the components of which may be seen separately with the naked eye by persons with very keen eyesight. If this star, however, be viewed with the telescope, the two companions will be seen far apart; and it will be noticed that each of them is again a double.

By means of the spectroscope Capella is shown to be really composed of two stars (one about twice as bright as the other) situated very close together and forming a binary system. Sirius is also a binary system; but it is what is called a "visual" one, for its component stars may be seen separately in very large telescopes. Its double, or rather binary, nature, was discovered in 1862 by the celebrated optician Alvan G. Clark, while in the act of testing the 18-inch refracting telescope, then just constructed by his firm, and now at the Dearborn Observatory, Illinois, U.S.A. The companion is only of the tenth magnitude, and revolves around Sirius in a period of about fifty years, at a mean distance equal to about that of Uranus from the sun. Seen from Sirius, it would shine only something like our full moon. It must be self-luminous and not a mere planet; for Mr. Gore has shown that if it shone only by the light reflected from Sirius, it would be quite invisible even in the Great Yerkes Telescope.

Procyon is also a binary, its companion having been discovered by Professor J.M. Schaeberle at the Lick Observatory in 1896. The period of revolution in this system is about forty years. Observations by Mr. T. Lewis of Greenwich seem, however, to point to the companion being a small nebula rather than a star.

The star (Eta) Cassiopeia (see Plate XIX., p. 292), is easily seen as a fine double in telescopes of moderate size. It is a binary system, the component bodies revolving around their common centre of gravity in a period of about two hundred years. This system is comparatively near to us, i.e. about nine light years, or a little further off than Sirius.

In a small telescope the star Castor will be found double, the components, one of which is brighter than the other, forming a binary system. The fainter of these was found by Belopolsky, with the spectroscope, to be composed of a system of two stars, one bright and the other either dark or not so bright, revolving around each other in a period of about three days. The brighter component of Castor is also a spectroscopic binary, with a period of about nine days; so that the whole of what we see with the naked eye as Castor, is in reality a remarkable system of four stars in mutual orbital movement.

Alpha Centauri--the nearest star to the earth--is a visual binary, the component bodies revolving around each other in a period of about eighty-one years. The extent of this system is about the same as that of Sirius. Viewed from each other, the bodies would shine only like our sun as seen from Neptune.

Among the numerous binary stars the orbits of some fifty have been satisfactorily determined. Many double stars, for which this has not yet been done, are, however, believed to be, without doubt, binary. In some cases a parallax has been found; so that we are enabled to estimate in miles the actual extent of such systems, and the masses of the bodies in terms of the sun's mass.

Most of the spectroscopic binaries appear to be upon a smaller scale than the telescopic ones. Some are, indeed, comparatively speaking, quite small. For instance, the component stars forming [b] Aurig?are about eight million miles apart, while in [z] Geminorum, the distance between the bodies is only a little more than a million miles.

Spectroscopic binaries are probably very numerous. Professor W.W. Campbell, Director of the Lick Observatory, estimates, for instance, that, out of about every half-a-dozen stars, one is a spectroscopic binary.

It is only in the case of binary systems that we can discover the masses of stars at all. These are ascertained from their movements with regard to each other under the influence of their mutual gravitative attractions. In the case of simple stars we have clearly nothing of the kind to judge by; though, if we can obtain a parallax, we may hazard a guess from their brightness.

Binary stars were incidentally discovered by Sir William Herschel. In his researches to get a stellar parallax he had selected a number of double stars for test purposes, on the assumption that, if one of such a pair were much nearer than the other, it might show a displacement with regard to its neighbour as a direct consequence of the earth's orbital movement around the sun. He, however, failed entirely to obtain any parallaxes, the triumph in this being, as we have seen, reserved for Bessel. But in some of the double stars which he had selected, he found certain alterations in the relative positions of the bodies, which plainly were not a consequence of the earth's motion, but showed rather that there was an actual circling movement of the bodies themselves under their mutual attractions. It is to be noted that the existence of such connected pairs had been foretold as probable by the Rev. John Michell, who lived a short time before Herschel.

The researches into binary systems--both those which can be seen with the eye and those which can be observed by means of the spectroscope, ought to impress upon us very forcibly the wide sway of the law of gravitation.

Of star clusters about 100 are known, and such systems often contain several thousand stars. They usually cover an area of sky somewhat smaller than the moon appears to fill. In most clusters the stars are very faint, and, as a rule, are between the twelfth and sixteenth magnitudes. It is difficult to say whether these are actually small bodies, or whether their faintness is due merely to their great distance from us, since they are much too far off to show any appreciable parallactic displacement. Mr. Gore, however, thinks there is good evidence to show that the stars in clusters are really close, and that the clusters themselves fill a comparatively small space.

One of the finest examples of a cluster is the great globular one, in the constellation of Hercules, discovered by Halley in 1714. It contains over 5000 stars, and upon a clear, dark night is visible to the naked eye as a patch of light. In the telescope, however, it is a wonderful object. There are also fine clusters in the constellations of Auriga, Pegasus, and Canes Venatici. In the southern heavens there are some magnificent examples of globular clusters. This hemisphere seems, indeed, to be richer in such objects than the northern. For instance, there is a great one in the constellation of the Centaur, containing some 6000 stars (see Plate XXI., p. 306).

Certain remarkable groups of stars, of a nature similar to clusters, though not containing such faint or densely packed stars as those we have just alluded to, call for a mention in this connection. The best example of such star groups are the Pleiades and the Hyades (see Plate XX., p. 296), Coma Berenices, and Prepe (or the Beehive), the last-named being in the constellation of Cancer.

Stars which alter in their brightness are called Variable Stars, or "variables." The first star whose variability attracted attention is that known as Omicron Ceti, namely, the star marked with the Greek letter [o] (Omicron) in the constellation of Cetus, or the Whale, a constellation situated not far from Taurus. This star, the variability of which was discovered by Fabricius in 1596, is also known as Mira, or the "Wonderful," on account of the extraordinary manner in which its light varies from time to time. The star known by the name of Algol,[32] popularly called the "Demon Star"--whose astronomical designation is [b] (Beta) Persei, or the star second in brightness in the constellation of Perseus--was discovered by Goodricke, in the year 1783, to be a variable star. In the following year [b] Lyr? the star in Lyra next in order of brightness after Vega, was also found by the same observer to be a variable. It may be of interest to the reader to know that Goodricke was deaf and dumb, and that he died in 1786 at the early age of twenty-one years!

It was not, however, until the close of the nineteenth century that much attention was paid to variable stars. Now several hundreds of these are known, thanks chiefly to the observations of, amongst others, Professor S.C. Chandler of Boston, U.S.A., Mr. John Ellard Gore of Dublin, and Dr. A.W. Roberts of South Africa. This branch of astronomy has not, indeed, attracted as much popular attention as it deserves, no doubt because the nature of the work required does not call for the glamour of an observatory or a large telescope.

The chief discoveries with regard to variable stars have been made by the naked eye, or with a small binocular. The amount of variation is estimated by a comparison with other stars. As in many other branches of astronomy, photography is now employed in this quest with marked success; and lately many variable stars have been found to exist in clusters and nebul?

It was at one time considered that a variable star was in all probability a body, a portion of whose surface had been relatively darkened in some manner akin to that in which sun spots mar the face of the sun; and that when its axial rotation brought the less illuminated portions in turn towards us, we witnessed a consequent diminution in the star's general brightness. Herschel, indeed, inclined to this explanation, for his belief was that all the stars bore spots like those of the sun. It appears preferably thought nowadays that disturbances take place periodically in the atmosphere or surroundings of certain stars, perhaps through the escape of imprisoned gases, and that this may be a fruitful cause of changes of brilliancy. The theory in question will, however, apparently account for only one class of variable star, namely, that of which Mira Ceti is the best-known example. The scale on which it varies in brightness is very great, for it changes from the second to the ninth magnitude. For the other leading type of variable star, Algol, of which mention has already been made, is the best instance. The shortness of the period in which the changes of brightness in such stars go their round, is the chief characteristic of this latter class. The period of Algol is a little under three days. This star when at its brightest is of about the second magnitude, and when least bright is reduced to below the third magnitude; from which it follows that its light, when at the minimum, is only about one-third of what it is when at the maximum. It seems definitely proved by means of the spectroscope that variables of this kind are merely binary stars, too close to be separated by the telescope, which, as a consequence of their orbits chancing to be edgewise towards us, eclipse each other in turn time after time. If, for instance, both components of such a pair are bright, then when one of them is right behind the other, we will not, of course, get the same amount of light as when they are side by side. If, on the other hand, one of the components happens to be dark or less luminous and the other bright, the manner in which the light of the bright star will be diminished when the darker star crosses its face should easily be understood. It is to the second of these types that Algol is supposed to belong. The Algol system appears to be composed of a body about as broad as our sun, which regularly eclipses a brighter body which has a diameter about half as great again.

Since the companion of Algol is often spoken of as a dark body, it were well here to point out that we have no evidence at all that it is entirely devoid of light. We have already found, in dealing with spectroscopic binaries, that when one of the component stars is below a certain magnitude[33] its spectrum will not be seen; so one is left in the glorious uncertainty as to whether the body in question is absolutely dark, or darkish, or faint, or indeed only just out of range of the spectroscope.

It is thought probable by good authorities that the companion of Algol is not quite dark, but has some inherent light of its own. It is, of course, much too near Algol to be seen with the largest telescope. There is in fact a distance of only from two to three millions of miles between the bodies, from which Mr. Gore infers that they would probably remain unseparated even in the largest telescope which could ever be constructed by man.

The number of known variables of the Algol type is, so far, small; not much indeed over thirty. In some of them the components are believed to revolve touching each other, or nearly so. An extreme example of this is found in the remarkable star V. Puppis, an Algol variable of the southern hemisphere. Both its components are bright, and the period of light variation is about one and a half days. Dr. A. W. Roberts finds that the bodies are revolving around each other in actual contact.

Temporary stars are stars which have suddenly blazed out in regions of the sky where no star was previously seen, and have faded away more or less gradually.

It was the appearance of such a star, in the year 134 B.C., which prompted Hipparchus to make his celebrated catalogue, with the object of leaving a record by which future observers could note celestial changes. In 1572 another star of this kind flashed out in the constellation of Cassiopeia (see Plate XIX., p. 292), and was detected by Tycho Brahe. It became as bright as the planet Venus, and eventually was visible in the day-time. Two years later, however, it disappeared, and has never since been seen. In 1604 Kepler recorded a similar star in the constellation of Ophiuchus which grew to be as bright as Jupiter. It also lasted for about two years, and then faded away, leaving no trace behind. It is rarely, however, that temporary stars attain to such a brilliance; and so possibly in former times a number of them may have appeared, but not have risen to a sufficient magnitude to attract attention. Even now, unless such a star becomes clearly visible to the naked eye, it runs a good chance of not being detected. A curious point, worth noting, with regard to temporary stars is that the majority of them have appeared in the Milky Way.

These sudden visitations have in our day received the name of Nov? that is to say, "New" Stars. Two, in recent years, attracted a good deal of attention. The first of these, known as Nova Aurig? or the New Star in the constellation of Auriga, was discovered by Dr. T.D. Anderson at Edinburgh in January 1892. At its greatest brightness it attained to about the fourth magnitude. By April it had sunk to the twelfth, but during August it recovered to the ninth magnitude. After this last flare-up it gradually faded away.

The startling suddenness with which temporary stars usually spring into being is the groundwork upon which theories to account for their origin have been erected. That numbers of dark stars, extinguished suns, so to speak, may exist in space, there is a strong suspicion; and it is just possible that we have an instance of one dark stellar body in the companion of Algol. That such dark stars might be in rapid motion is reasonable to assume from the already known movements of bright stars. Two dark bodies might, indeed, collide together, or a collision might take place between a dark star and a star too faint to be seen even with the most powerful telescope. The conflagration produced by the impact would thus appear where nothing had been seen previously. Again, a similar effect might be produced by a dark body, or a star too faint to be seen, being heated to incandescence by plunging in its course through a nebulous mass of matter, of which there are many examples lying about in space.

The last explanation, which is strongly reminiscent of what takes place in shooting stars, appears more probable than the collision theory. The flare-up of new stars continues, indeed, only for a comparatively short time; whereas a collision between two bodies would, on the other hand, produce an enormous nebula which might take even millions of years to cool down. We have, indeed, no record of any such sudden appearance of a lasting nebula.

The other temporary star, known as Nova Persei, or the new star in the constellation of Perseus, was discovered early in the morning of February 22, 1901, also by Dr. Anderson. A day later it had grown to be brighter than Capella. Photographs which had been taken, some three days previous to its discovery, of the very region of the sky in which it had burst forth, were carefully examined, and it was not found in these. At the end of two days after its discovery Nova Persei had lost one-third of its light. During the ensuing six months it passed through a series of remarkable fluctuations, varying in brightness between the third and fifth magnitudes. In the month of August it was seen to be surrounded by luminous matter in the form of a nebula, which appeared to be gradually spreading to some distance around. Taking into consideration the great way off at which all this was taking place, it looked as if the new star had ejected matter which was travelling outward with a velocity equivalent to that of light. The remarkable theory was, however, put forward by Professor Kapteyn and the late Dr. W.E. Wilson that there might be after all no actual transmission of matter; but that perhaps the real explanation was the gradual illumination of hitherto invisible nebulous matter, as a consequence of the flare-up which had taken place about six months before. It was, therefore, imagined that some dark body moving through space at a very rapid rate had plunged through a mass of invisible nebulous matter, and had consequently become heated to incandescence in its passage, very much like what happens to a meteor when moving through our atmosphere. The illumination thus set up temporarily in one point, being transmitted through the nebulous wastes around with the ordinary velocity of light, had gradually rendered this surrounding matter visible. On the assumptions required to fit in with such a theory, it was shown that Nova Persei must be at a distance from which light would take about three hundred years in coming to us. The actual outburst of illumination, which gave rise to this temporary star, would therefore have taken place about the beginning of the reign of James I.

Some recent investigations with regard to Nova Persei have, however, greatly narrowed down the above estimate of its distance from us. For instance, Bergstrand proposes a distance of about ninety-nine light years; while the conclusions of Mr. F.W. Very would bring it still nearer, i.e. about sixty-five light years.

The last celestial objects with which we have here to deal are the Nebul? These are masses of diffused shining matter scattered here and there through the depths of space. Nebul?are of several kinds, and have been classified under the various headings of Spiral, Planetary, Ring, and Irregular.

A typical spiral nebula is composed of a disc-shaped central portion, with long curved arms projecting from opposite sides of it, which give an impression of rapid rotatory movement.

The discovery of spiral nebul?was made by Lord Rosse with his great 6-foot reflector. Two good examples of these objects will be found in Ursa Major, while there is another fine one in Canes Venatici (see Plate XXII., p. 314), a constellation which lies between Ursa Major and Boes. But the finest spiral of all, perhaps the most remarkable nebula known to us, is the Great Nebula in the constellation of Andromeda, (see Plate XXIII., p. 316)--a constellation just further from the pole than Cassiopeia. When the moon is absent and the night clear this nebula can be easily seen with the naked eye as a small patch of hazy light. It is referred to by Al Sufi.

[Illustration: PLATE XXII. SPIRAL NEBULA IN THE CONSTELLATION OF CANES VENATICI

From a photograph by the late Dr. W.E. Wilson, D.Sc., F.R.S.

(Page 314)]

Spiral nebul?are white in colour, whereas the other kinds of nebula have a greenish tinge. They are also by far the most numerous; and the late Professor Keeler, who considered this the normal type of nebula, estimated that there were at least 120,000 of such spirals within the reach of the Crossley reflector of the Lick Observatory. Professor Perrine has indeed lately raised this estimate to half a million, and thinks that with more sensitive photographic plates and longer exposures the number of spirals would exceed a million. The majority of these objects are very small, and appear to be distributed over the sky in a fairly uniform manner.

Planetary nebul?are small faint roundish objects which, when seen in the telescope, recall the appearance of a planet, hence their name. One of these nebul? known astronomically as G.C. 4373, has recently been found to be rushing through space towards the earth at a rate of between thirty and forty miles per second. It seems strange, indeed, that any gaseous mass should move at such a speed!

What are known as ring nebul?were until recently believed to form a special class. These objects have the appearance of mere rings of nebulous matter. Much doubt has, however, been thrown upon their being rings at all; and the best authorities regard them merely as spiral nebul? of which we happen to get a foreshortened view. Very few examples are known, the most famous being one in the constellation of Lyra, usually known as the Annular Nebula in Lyra. This object is so remote from us as to be entirely invisible to the naked eye. It contains a star of the fifteenth magnitude near to its centre. From photographs taken with the Crossley reflector, Professor Schaeberle finds in this nebula evidences of spiral structure. It may here be mentioned that the Great Nebula in Andromeda, which has now turned out to be a spiral, had in earlier photographs the appearance of a ring.

There also exist nebul?of irregular form, the most notable being the Great Nebula in the constellation of Orion (see Plate XXIV., p. 318). It is situated in the centre of the "Sword" of Orion (see Plate XX., p. 296). In large telescopes it appears as a magnificent object, and in actual dimensions it must be much on the same scale as the Andromeda Nebula. The spectroscope tells us that it is a mass of glowing gas.

The Trifid Nebula, situated in the constellation of Sagittarius, is an object of very strange shape. Three dark clefts radiate from its centre, giving it an appearance as if it had been torn into shreds.

The Dumb-bell Nebula, a celebrated object, so called from its likeness to a dumb-bell, turns out, from recent photographs taken by Professor Schaeberle, which bring additional detail into view, to be after all a great spiral.

There is a nest, or rather a cluster of nebul?in the constellation of Coma Berenices; over a hundred of these objects being here gathered into a space of sky about the size of our full moon.

The spectroscope informs us that spiral nebul?are composed of partially-cooled matter. Their colour, as we have seen, is white. Nebul?of a greenish tint are, on the other hand, found to be entirely in a gaseous condition. Just as the solar corona contains an unknown element, which for the time being has been called "Coronium," so do the gaseous nebul?give evidence of the presence of another unknown element. To this Sir William Huggins has given the provisional name of "Nebulium."

The Magellanic Clouds are two patches of nebulous-looking light, more or less circular in form, which are situated in the southern hemisphere of the sky. They bear a certain resemblance to portions of the Milky Way, but are, however, not connected with it. They have received their name from the celebrated navigator, Magellan, who seems to have been one of the first persons to draw attention to them. "Nubecul? is another name by which they are known, the larger cloud being styled nubecula major and the smaller one nubecula minor. They contain within them stars, clusters, and gaseous nebul? No parallax has yet been found for any object which forms part of the nubecul? so it is very difficult to estimate at what distance from us they may lie. They are, however, considered to be well within our stellar universe.

Having thus brought to a conclusion our all too brief review of the stars and the nebul?-of the leading objects in fine which the celestial spaces have revealed to man--we will close this chapter with a recent summation by Sir David Gill of the relations which appear to obtain between these various bodies. "Huggins's spectroscope," he says, "has shown that many nebul?are not stars at all; that many well-condensed nebul? as well as vast patches of nebulous light in the sky, are but inchoate masses of luminous gas. Evidence upon evidence has accumulated to show that such nebul?consist of the matter out of which stars (i.e. suns) have been and are being evolved. The different types of star spectra form such a complete and gradual sequence (from simple spectra resembling those of nebul?onwards through types of gradually increasing complexity) as to suggest that we have before us, written in the cryptograms of these spectra, the complete story of the evolution of suns from the inchoate nebula onwards to the most active sun (like our own), and then downward to the almost heatless and invisible ball. The period during which human life has existed upon our globe is probably too short-- even if our first parents had begun the work--to afford observational proof of such a cycle of change in any particular star; but the fact of such evolution, with the evidence before us, can hardly be doubted."[34]

[32] The name Al go, meaning the Demon, was what the old Arabian astronomers called it, which looks very much as if they had already noticed its rapid fluctuations in brightness.

[33] Mr. Gore thinks that the companion of Algol may be a star of the sixth magnitude.

[34] Presidential Address to the British Association for the Advancement of Science (Leicester, 1907), by Sir David Gill, K.C.B., LL.D., F.R.S., &c. &c.

CHAPTER XXV

THE STELLAR UNIVERSE

The stars appear fairly evenly distributed all around us, except in one portion of the sky where they seem very crowded, and so give one an impression of being very distant. This portion, known as the Milky Way, stretches, as we have already said, in the form of a broad band right round the entire heavens. In those regions of the sky most distant from the Milky Way the stars appear to be thinly sown, but become more and more closely massed together as the Milky Way is approached.

This apparent distribution of the stars in space has given rise to a theory which was much favoured by Sir William Herschel, and which is usually credited to him, although it was really suggested by one Thomas Wright of Durham in 1750; that is to say, some thirty years or more before Herschel propounded it. According to this, which is known as the "Disc" or "Grindstone" Theory, the stars are considered as arranged in space somewhat in the form of a thick disc, or grindstone, close to the central parts of which our solar system is situated.[35] Thus we should see a greater number of stars when we looked out through the length of such a disc in any direction, than when we looked out through its breadth. This theory was, for a time, supposed to account quite reasonably for the Milky Way, and for the gradual increase in the number of stars in its vicinity.

It is quite impossible to verify directly such a theory, for we know the actual distance of only about forty-three stars. We are unable, therefore, definitely to assure ourselves whether, as the grindstone theory presupposes, the stellar universe actually reaches out very much further from us in the direction of the Milky Way than in the other parts of the sky. The theory is clearly founded upon the supposition that the stars are more or less equal in size, and are scattered through space at fairly regular distances from each other.

Brightness, therefore, had been taken as implying nearness to us, and faintness great distance. But we know to-day that this is not the case, and that the stars around us are, on the other hand, of various degrees of brightness and of all orders of size. Some of the faint stars--for instance, the galloping star in Pictor--are indeed nearer to us than many of the brighter ones. Sirius, on the other hand, is twice as far off from us as [a] Centauri, and yet it is very much brighter; while Canopus, which in brightness is second only to Sirius out of the whole sky, is too far off for its distance to be ascertained! It must be remembered that no parallax had yet been found for any star in the days of Herschel, and so his estimations of stellar distances were necessarily of a very circumstantial kind. He did not, however, continue always to build upon such uncertain ground; but, after some further examination of the Milky Way, he gave up his idea that the stars were equally disposed in space, and eventually abandoned the grindstone theory.

Since we have no means of satisfactorily testing the matter, through finding out the various distances from us at which the stars are really placed, one might just as well go to the other extreme, and assume that the thickening of stars in the region of the Milky Way is not an effect of perspective at all, but that the stars in that part of the sky are actually more crowded together than elsewhere--a thing which astronomers now believe to be the case. Looked at in this way, the shape of the stellar universe might be that of a globe-shaped aggregation of stars, in which the individuals are set at fairly regular distances from each other; the whole being closely encircled by a belt of densely packed stars. It must, however, be allowed that the gradual increase in the number of stars towards the Milky Way appears a strong argument in favour of the grindstone theory; yet the belt theory, as above detailed, seems to meet with more acceptance.

There is, in fact, one marked circumstance which is remarkably difficult of explanation by means of the grindstone theory. This is the existence of vacant spaces--holes, so to speak, in the groundwork of the Milky Way. For instance, there is a cleft running for a good distance along its length, and there is also a starless gap in its southern portion. It seems rather improbable that such a great number of stars could have arranged themselves so conveniently, as to give us a clear view right out into empty space through such a system in its greatest thickness; as if, in fact, holes had been bored, and clefts made, from the boundary of the disc clean up to where our solar system lies. Sir John Herschel long ago drew attention to this point very forcibly. It is plain that such vacant spaces can, on the other hand, be more simply explained as mere holes in a belt; and the best authorities maintain that the appearance of the Milky Way confirms a view of this kind.

Whichever theory be indeed the correct one, it appears at any rate that the stars do not stretch out in every direction to an infinite distance; but that the stellar system is of limited extent, and has in fact a boundary.

In the first place, Science has no grounds for supposing that light is in any way absorbed or destroyed merely by its passage through the "ether," that imponderable medium which is believed to transmit the luminous radiations through space. This of course is tantamount to saying that all the direct light from all the stars should reach us, excepting that little which is absorbed in its passage through our own atmosphere. If stars, and stars, and stars existed in every direction outwards without end, it can be proved mathematically that in such circumstances there could not remain the tiniest space in the sky without a star to fill it, and that therefore the heavens would always blaze with light, and the night would be as bright as the noonday.[36] How very far indeed this is from being the case, may be gathered from an estimate which has been made of the general amount of light which we receive from the stars. According to this estimate the sky is considered as more or less dark, the combined illumination sent to us by all the stars being only about the one-hundreth part of what we get from the full moon.[37]

Secondly, it has been suggested that although light may not suffer any extinction or diminution from the ether itself, still a great deal of illumination may be prevented from reaching us through myriads of extinguished suns, or dark meteoric matter lying about in space. The idea of such extinguished suns, dark stars in fact, seems however to be merely founded upon the sole instance of the invisible companion of Algol; but, as we have seen, there is no proof whatever that it is a dark body. Again, some astronomers have thought that the dark holes in the Milky Way, "Coal Sacks," as they are called, are due to masses of cool, or partially cooled matter, which cuts off the light of the stars beyond. The most remarkable of these holes is one in the neighbourhood of the Southern Cross, known as the "Coal Sack in Crux." But Mr. Gore thinks that the cause of the holes is to be sought for rather in what Sir William Herschel termed "clustering power," i.e. a tendency on the part of stars to accumulate in certain places, thus leaving others vacant; and the fact that globular and other clusters are to be found very near to such holes certainly seems corroborative of this theory. In summing up the whole question, Professor Newcomb maintains that there does not appear any evidence of the light from the Milky Way stars, which are apparently the furthest bodies we see, being intercepted by dark bodies or dark matter. As far as our telescopes can penetrate, he holds that we see the stars just as they are.

Also, if there did exist an infinite number of stars, one would expect to find evidence in some direction of an overpoweringly great force,--the centre of gravity of all these bodies.

It is noticed, too, that although the stars increase in number with decrease in magnitude, so that as we descend in the scale we find three times as many stars in each magnitude as in the one immediately above it, yet this progression does not go on after a while. There is, in fact, a rapid falling off in numbers below the twelfth magnitude; which looks as if, at a certain distance from us, the stellar universe were beginning to thin out.

Again, it is estimated, by Mr. Gore and others, that only about 100 millions of stars are to be seen in the whole of the sky with the best optical aids. This shows well the limited extent of the stellar system, for the number is not really great. For instance, there are from fifteen to sixteen times as many persons alive upon the earth at this moment!

Last of all, there appears to be strong photographic evidence that our sidereal system is limited in extent. Two photographs taken by the late Dr. Isaac Roberts of a region rich in stellar objects in the constellation of Cygnus, clearly show what has been so eloquently called the "darkness behind the stars." One of these photographs was taken in 1895, and the other in 1898. On both occasions the state of the atmosphere was practically the same, and the sensitiveness of the films was of the same degree. The exposure in the first case was only one hour; in the second it was about two hours and a half. And yet both photographs show exactly the same stars, even down to the faintest. From this one would gather that the region in question, which is one of the most thickly star-strewn in the Milky Way, is penetrable right through with the means at our command. Dr. Roberts himself in commenting upon the matter drew attention to the fact, that many astronomers seemed to have tacitly adopted the assumption that the stars extend indefinitely through space.

From considerations such as these the foremost astronomical authorities of our time consider themselves justified in believing that the collection of stars around us is finite; and that although our best telescopes may not yet be powerful enough to penetrate to the final stars, still the rapid decrease in numbers as space is sounded with increasing telescopic power, points strongly to the conclusion that the boundaries of the stellar system may not lie very far beyond the uttermost to which we can at present see.

Is it possible then to make an estimate of the extent of this stellar system?

Whatever estimates we may attempt to form cannot however be regarded as at all exact, for we know the actual distances of such a very few only of the nearest of the stars. But our knowledge of the distances even of these few, permits us to assume that the stars close around us may be situated, on an average, at about eight light-years from each other; and that this holds good of the stellar spaces, with the exception of the encircling girdle of the Milky Way, where the stars seem actually to be more closely packed together. This girdle further appears to contain the greater number of the stars. Arguing along these lines, Professor Newcomb reaches the conclusion that the farthest stellar bodies which we see are situated at about between 3000 and 4000 light-years from us.

Starting our inquiry from another direction, we can try to form an estimate by considering the question of proper motions.

It will be noticed that such motions do not depend entirely upon the actual speed of the stars themselves, but that some of the apparent movement arises indirectly from the speed of our own sun. The part in a proper motion which can be ascribed to the movement of our solar system through space is clearly a displacement in the nature of a parallax--Sir William Herschel called it "Systematic Parallax"; so that knowing the distance which we move over in a certain lapse of time, we are able to hazard a guess at the distances of a good many of the stars. An inquiry upon such lines must needs be very rough, and is plainly based upon the assumption that the stars whose distances we attempt to estimate are moving at an average speed much like that of our own sun, and that they are not "runaway stars" of the 1830 Groombridge order. Be that as it may, the results arrived at by Professor Newcomb from this method of reasoning are curiously enough very much on a par with those founded on the few parallaxes which we are really certain about; with the exception that they point to somewhat closer intervals between the individual stars, and so tend to narrow down our previous estimate of the extent of the stellar system.

Thus far we get, and no farther. Our solar system appears to lie somewhere near the centre of a great collection of stars, separated each one from the other, on an average, by some 40 billions of miles; the whole being arranged in the form of a mighty globular cluster. Light from the nearest of these stars takes some four years to come to us. It takes about 1000 times as long to reach us from the confines of the system. This globe of stars is wrapt around closely by a stellar girdle, the individual stars in which are set together more densely than those in the globe itself. The entire arrangement appears to be constructed upon a very regular plan. Here and there, as Professor Newcomb points out, the aspect of the heavens differs in small detail; but generally it may be laid down that the opposite portions of the sky, whether in the Milky Way itself, or in those regions distant from it, show a marked degree of symmetry. The proper motions of stars in corresponding portions of the sky reveal the same kind of harmony, a harmony which may even be extended to the various colours of the stars. The stellar system, which we see disposed all around us, appears in fine to bear all the marks of an organised whole.

The older astronomers, to take Sir William Herschel as an example, supposed some of the nebul?to be distant "universes." Sir William was led to this conclusion by the idea he had formed that, when his telescopes failed to show the separate stars of which he imagined these objects to be composed, he must put down the failure to their stupendous distance from us. For instance, he thought the Orion Nebula, which is now known to be made up of glowing gas, to be an external stellar system. Later on, however, he changed his mind upon this point, and came to the conclusion that "shining fluid" would better account both for this nebula, and for others which his telescopes had failed to separate into component stars.

The old ideas with regard to external systems and distant universes have been shelved as a consequence of recent research. All known clusters and nebul?are now firmly believed to lie within our stellar system.

This view of the universe of stars as a sort of island in the immensities, does not, however, give us the least idea about the actual extent of space itself. Whether what is called space is really infinite, that is to say, stretches out unendingly in every direction, or whether it has eventually a boundary somewhere, are alike questions which the human mind seems utterly unable to picture to itself.

[35] The Ptolemaic idea dies hard!

[36] Even the Milky Way itself is far from being a blaze of light, which shows that the stars composing it do not extend outwards indefinitely.

[37] Mr. Gore has recently made some remarkable deductions, with regard to the amount of light which we get from the stars. He considers that most of this light comes from stars below the sixth magnitude; and consequently, if all the stars visible to the naked eye were to be blotted out, the glow of the night sky would remain practically the same as it is at present. Going to the other end of the scale, he thinks also that the combined light which we get from all the stars below the seventeenth magnitude is so very small, that it may be neglected in such an estimation. He finds, indeed, that if there are stars so low as the twentieth magnitude, one hundred millions of them would only be equal in brightness to a single first-magnitude star like Vega. On the other hand, it is possible that the light of the sky at night is not entirely due to starlight, but that some of it may be caused by phosphorescent glow.

CHAPTER XXVI

THE STELLAR UNIVERSE--continued

It is very interesting to consider the proper motions of stars with reference to such an isolated stellar system as has been pictured in the previous chapter. These proper motions are so minute as a rule, that we are quite unable to determine whether the stars which show them are moving along in straight lines, or in orbits of immense extent. It would, in fact, take thousands of years of careful observation to determine whether the paths in question showed any degree of curving. In the case of the more distant stars, the accurate observations which have been conducted during the last hundred years have not so far revealed any proper motions with regard to them; but one cannot escape the conclusion that these stars move as the others do.

If space outside our stellar system is infinite in extent, and if all the stars within that system are moving unchecked in every conceivable direction, the result must happen that after immense ages these stars will have drawn apart to such a distance from each other, that the system will have entirely disintegrated, and will cease to exist as a connected whole. Eventually, indeed, as Professor Newcomb points out, the stars will have separated so far from each other that each will be left by itself in the midst of a black and starless sky. If, however, a certain proportion of stars have a speed sufficiently slow, they will tend under mutual attraction to be brought to rest by collisions, or forced to move in orbits around each other. But those stars which move at excessive speeds, such, for instance, as 1830 Groombridge, or the star in the southern constellation of Pictor, seem utterly incapable of being held back in their courses by even the entire gravitative force of our stellar system acting as a whole. These stars must, therefore, move eventually right through the system and pass out again into the empty spaces beyond. Add to this; certain investigations, made into the speed of 1830 Groombridge, furnish a remarkable result. It is calculated, indeed, that had this star been falling through infinite space for ever, pulled towards us by the combined gravitative force of our entire system of stars, it could not have gathered up anything like the speed with which it is at present moving. No force, therefore, which we can conjure out of our visible universe, seems powerful enough either to have impressed upon this runaway star the motion which it now has, or to stay it in its wild course. What an astounding condition of things!

Speculations like this call up a suspicion that there may yet exist other universes, other centres of force, notwithstanding the apparent solitude of our stellar system in space. It will be recollected that the idea of this isolation is founded upon such facts as, that the heavens do not blaze with light, and that the stars gradually appear to thin out as we penetrate the system with increasing telescopic power. But perchance there is something which hinders us from seeing out into space beyond our cluster of stars; which prevents light, in fact, from reaching us from other possible systems scattered through the depths beyond. It has, indeed, been suggested by Mr. Gore[38] that the light-transmitting ether may be after all merely a kind of "atmosphere" of the stars; and that it may, therefore, thin off and cease a little beyond the confines of our stellar system, just as the air thins off and practically ceases at a comparatively short distance from the earth. A clashing together of solid bodies outside our atmosphere could plainly send us no sound, for there is no air extending the whole way to bear to our ears the vibrations thus set up; so light emitted from any body lying beyond our system of stars, would not be able to come to us if the ether, whose function it is to convey the rays of light, ceased at or near the confines of that system.

Perchance we have in this suggestion the key to the mystery of how our sun and the other stellar bodies maintain their functions of temperature and illumination. The radiations of heat and light arriving at the limits of this ether, and unable to pass any further, may be thrown back again into the system in some altered form of energy.

But these, at best, are mere airy and fascinating speculations. We have, indeed, no evidence whatever that the luminiferous ether ceases at the boundary of the stellar system. If, therefore, it extends outwards infinitely in every direction, and if it has no absorbing or weakening effect on the vibrations which it transmits, we cannot escape from the conclusion that practically all the rays of light ever emitted by all the stars must chase one another eternally through the never-ending abysses of space.

[38] Planetary and Stellar Studies, by John Ellard Gore, F.R.A.S., M.R.I.A., London, 1888.

CHAPTER XXVII

THE BEGINNING OF THINGS

LAPLACE'S NEBULAR HYPOTHESIS

Dwelling upon the fact that all the motions of revolution and rotation in the solar system, as known in his day, took place in the same direction and nearly in the same plane, the great French astronomer, Laplace, about the year 1796, put forward a theory to account for the origin and evolution of that system. He conceived that it had come into being as a result of the gradual contraction, through cooling, of an intensely heated gaseous lens-shaped mass, which had originally occupied its place, and had extended outwards beyond the orbit of the furthest planet. He did not, however, attempt to explain how such a mass might have originated! He went on to suppose that this mass, in some manner, perhaps by mutual gravitation among its parts, had acquired a motion of rotation in the same direction as the planets now revolve. As this nebulous mass parted with its heat by radiation, it contracted towards the centre. Becoming smaller and smaller, it was obliged to rotate faster and faster in order to preserve its equilibrium. Meanwhile, in the course of contraction, rings of matter became separated from the nucleus of the mass, and were left behind at various intervals. These rings were swept up into subordinate masses similar to the original nebula. These subordinate masses also contracted in the same manner, leaving rings behind them which, in turn, were swept up to form satellites. Saturn's ring was considered, by Laplace, as the only portion of the system left which still showed traces of this evolutionary process. It is even probable that it may have suggested the whole of the idea to him.

Laplace was, however, not the first philosopher who had speculated along these lines concerning the origin of the world.

Nearly fifty years before, in 1750 to be exact, Thomas Wright, of Durham, had put forward a theory to account for the origin of the whole sidereal universe. In his theory, however, the birth of our solar system was treated merely as an incident. Shortly afterwards the subject was taken up by the famous German philosopher, Kant, who dealt with the question in a still more ambitious manner, and endeavoured to account in detail for the origin of the solar system as well as of the sidereal universe. Something of the trend of such theories may be gathered from the remarkable lines in Tennyson's Princess:--

"This world was once a fluid haze of light, Till toward the centre set the starry tides, And eddied into suns, that wheeling cast The planets."

The theory, as worked out by Kant, was, however, at the best merely a tour de force of philosophy. Laplace's conception was much less ambitious, for it did not attempt to explain the origin of the entire universe, but only of the solar system. Being thus reasonably limited in its scope, it more easily obtained credence. The arguments of Laplace were further founded upon a mathematical basis. The great place which he occupied among the astronomers of that time caused his theory to exert a preponderating influence on scientific thought during the century which followed.

A modification of Laplace's theory is the Meteoritic Hypothesis of Sir Norman Lockyer. According to the views of that astronomer, the material of which the original nebula was composed is presumed to have been in the meteoric, rather than in the gaseous, state. Sir Norman Lockyer holds, indeed, that nebul?are, in reality, vast swarms of meteors, and the light they emit results from continual collisions between the constituent particles. The French astronomer, Faye, also proposed to modify Laplace's theory by assuming that the nebula broke up into rings all at once, and not in detail, as

Laplace had wished to suppose.

The hypothesis of Laplace fits in remarkably well with the theory put forward in later times by Helmholtz, that the heat of the sun is kept up by the continual contraction of its mass. It could thus have only contracted to its present size from one very much larger.

Plausible, however, as Laplace's great hypothesis appears on the surface, closer examination shows several vital objections, a few of those set forth by Professor Moulton being here enumerated--

Although Laplace held that the orbits of the planets were sufficiently near to being in the one plane to support his views, yet later investigators consider that their very deviations from this plane are a strong argument against the hypothesis.

Again, it is thought that if the theory were the correct explanation, the various orbits of the planets would be much more nearly circular than they are.

It is also thought that such interlaced paths, as those in which the asteroids and the little planet Eros move, are most unlikely to have been produced as a result of Laplace's nebula.

Further, while each of the rings was sweeping up its matter into a body of respectable dimensions, its gravitative power would have been for the time being so weak, through being thus spread out, that any lighter elements, as, for instance, those of the gaseous order, would have escaped into space in accordance with the principles of the kinetic theory.

The idea that rings would at all be left behind at certain intervals during the contraction of the nebula is, perhaps, one of the weakest points in Laplace's hypothesis.

Mathematical investigation does not go to show that the rings, presuming they could be left behind during the contraction of the mass, would have aggregated into planetary bodies. Indeed, it rather points to the reverse.

Lastly, such a discovery as that the ninth satellite of Saturn revolves in a retrograde direction--that is to say, in a direction contrary to the other revolutions and rotations in our solar system--appears directly to contradict the hypothesis.

Although Laplace's hypothesis seems to break down under the keen criticism to which it has been subjected, yet astronomers have not relinquished the idea that our solar system has probably had its origin from a nebulous mass. But the apparent failure of the Laplacian theory is emphasised by the fact, that not a single example of a nebula, in the course of breaking up into concentric rings, is known to exist in the entire heaven. Indeed, as we saw in Chapter XXIV., there seems to be no reliable example of even a "ring" nebula at all. Mr. Gore has pointed this out very succinctly in his recently published work, Astronomical Essays, where he says:--"To any one who still persists in maintaining the hypothesis of ring formation in nebul? it may be said that the whole heavens are against him."

The conclusions of Keeler already alluded to, that the spiral is the normal type of nebula, has led during the past few years to a new theory by the American astronomers, Professors Chamberlin and Moulton. In the detailed account of it which they have set forth, they show that those anomalies which were stumbling-blocks to Laplace's theory do not contradict theirs. To deal at length with this theory, to which the name of "Planetesimal Hypothesis" has been given, would not be possible in a book of this kind. But it may be of interest to mention that the authors of the theory in question remount the stream of time still further than did Laplace, and seek to explain the origin of the spiral nebul?themselves in the following manner:--

Having begun by assuming that the stars are moving apparently in every direction with great velocities, they proceed to point out that sooner or later, although the lapse of time may be extraordinarily long, collisions or near approaches between stars are bound to occur. In the case of collisions the chances are against the bodies striking together centrally, it being very much more likely that they will hit each other rather towards the side. The nebulous mass formed as a result of the disintegration of the bodies through their furious impact would thus come into being with a spinning movement, and a spiral would ensue. Again, the stars may not actually collide, but merely approach near to each other. If very close, the interaction of gravitation will give rise to intense strains, or tides, which will entirely disintegrate the bodies, and a spiral nebula will similarly result. As happens upon our earth, two such tides would rise opposite to each other; and, consequently, it is a noticeable fact that spiral nebul?have almost invariably two opposite branches (see Plate XXII., p 314). Even if not so close, the gravitational strains set up would produce tremendous eruptions of matter; and in this case, a spiral movement would also be generated. On such an assumption the various bodies of the solar system may be regarded as having been ejected from parent masses.

The acceptance of the Planetesimal Hypothesis in the place of the Hypothesis of Laplace will not, as we have seen, by any means do away with the probability that our solar system, and similar systems, have originated from a nebulous mass. On the contrary it puts that idea on a firmer footing than before. The spiral nebul?which we see in the heavens are on a vast scale, and may represent the formation of stellar systems and globular clusters. Our solar system may have arisen from a small spiral.

We will close these speculations concerning the origin of things with a short sketch of certain investigations made in recent years by Sir George H. Darwin, of Cambridge University, into the question of the probable birth of our moon. He comes to the conclusion that at least fifty-four millions of years ago the earth and moon formed one body, which had a diameter of a little over 8000 miles. This body rotated on an axis in about five hours, namely, about five times as fast as it does at present. The rapidity of the rotation caused such a tremendous strain that the mass was in a condition of, what is called, unstable equilibrium; very little more, in fact, being required to rend it asunder. The gravitational pull of the sun, which, as we have already seen, is in part the cause of our ordinary tides, supplied this extra strain, and a portion of the mass consequently broke off, which receded gradually from the rest and became what we now know as the moon. Sir George Darwin holds that the gravitational action of the sun will in time succeed in also disturbing the present apparent harmony of the earth-moon system, and will eventually bring the moon back towards the earth, so that after the lapse of great ages they will re-unite once again.

In support of this theory of the terrestrial origin of the moon, Professor W.H. Pickering has put forward a bold hypothesis that our satellite had its origin in the great basin of the Pacific. This ocean is roughly circular, and contains no large land masses, except the Australian Continent. He supposes that, prior to the moon's birth, our globe was already covered with a slight crust. In the tearing away of that portion which was afterwards destined to become the moon the remaining area of the crust was rent in twain by the shock; and thus were formed the two great continental masses of the Old and New Worlds. These masses floated apart across the fiery ocean, and at last settled in the positions which they now occupy. In this way Professor Pickering explains the remarkable parallelism which exists between the opposite shores of the Atlantic. The fact of this parallelism had, however, been noticed before; as, for example, by the late Rev. S.J. Johnson, in his book Eclipses, Past and Future, where we find the following passage:--

"If we look at our maps we shall see the parts of one Continent that jut out agree with the indented portions of another. The prominent coast of Africa would fit in the opposite opening between North and South America, and so in numerous other instances. A general rending asunder of the World would seem to have taken place when the foundations of the great deep were broken up."

Although Professor Pickering's theory is to a certain degree anticipated in the above words, still he has worked out the idea much more fully, and given it an additional fascination by connecting it with the birth of the moon. He points out, in fact, that there is a remarkable similarity between the lunar volcanoes and those in the immediate neighbourhood of the Pacific Ocean. He goes even further to suggest that Australia is another portion of the primal crust which was detached out of the region now occupied by the Indian Ocean, where it was originally connected with the south of India or the east of Africa.

Certain objections to the theory have been put forward, one of which is that the parallelism noticed between the opposite shores of the Atlantic is almost too perfect to have remained through some sixty millions of years down to our own day, in the face of all those geological movements of upheaval and submergence, which are perpetually at work upon our globe. Professor Pickering, however, replies to this objection by stating that many geologists believe that the main divisions of land and water on the earth are permanent, and that the geological alterations which have taken place since these were formed have been merely of a temporary and superficial nature.

CHAPTER XXVIII

THE END OF THINGS

We have been trying to picture the beginning of things. We will now try to picture the end.

In attempting this, we find that our theories must of necessity be limited to the earth, or at most to the solar system. The time-honoured expression "End of the World" really applies to very little beyond the end of our own earth. To the people of past ages it, of course, meant very much more. For them, as we have seen, the earth was the centre of everything; and the heavens and all around were merely a kind of minor accompaniment, created, as they no doubt thought, for their especial benefit. In the ancient view, therefore, the beginning of the earth meant the beginning of the universe, and the end of the earth the extinction of all things. The belief, too, was general that this end would be accomplished through fire. In the modern view, however, the birth and death of the earth, or indeed of the solar system, might pass as incidents almost unnoticed in space. They would be but mere links in the chain of cosmic happenings.

A number of theories have been forward from time to time prognosticating the end of the earth, and consequently of human life. We will conclude with a recital of a few of them, though which, if any, is the true one, the Last Men alone can know.

Just as a living creature may at any moment die in the fulness of strength through sudden malady or accident, or, on the other hand, may meet with death as a mere consequence of old age, so may our globe be destroyed by some sudden cataclysm, or end in slow processes of decay. Barring accidents, therefore, it would seem probable that the growing cold of the earth, or the gradual extinction of the sun, should after many millions of years close the chapter of life, as we know it. On the former of these suppositions, the decrease of temperature on our globe might perhaps be accelerated by the thinning of the atmosphere, through the slow escape into space of its constituent gases, or their gradual chemical combination with the materials of the earth. The subterranean heat entirely radiated away, there would no longer remain any of those volcanic elevating forces which so far have counteracted the slow wearing down of the land surface of our planet, and thus what water remained would in time wash over all. If this preceded the growing cold of the sun, certain strange evolutions of marine forms of life would be the last to endure, but these, too, would have to go in the end.

Should, however, the actual process be the reverse of this, and the sun cool down the quicker, then man would, as a consequence of his scientific knowledge, tend in all probability to outlive the other forms of terrestrial life. In such a vista we can picture the regions of the earth towards the north and south becoming gradually more and more uninhabitable through cold, and human beings withdrawing before the slow march of the icy boundary, until the only regions capable of habitation would lie within the tropics. In such a struggle between man and destiny science would be pressed to the uttermost, in the devising of means to counteract the slow diminution of the solar heat and the gradual disappearance of air and water. By that time the axial rotation of our globe might possibly have been slowed down to such an extent that one side alone of its surface would be turned ever towards the fast dying sun. And the mind's eye can picture the last survivors of the human race, huddled together for warmth in a glass-house somewhere on the equator, waiting for the end to come.

The mere idea of the decay and death of the solar system almost brings to one a cold shudder. All that sun's light and heat, which means so much to us, entirely a thing of the past. A dark, cold ball rushing along in space, accompanied by several dark, cold balls circling ceaselessly around it. One of these a mere cemetery, in which there would be no longer any recollection of the mighty empires, the loves and hates, and all that teeming play of life which we call History. Tombstones of men and of deeds, whirling along forgotten in the darkness and silence. Sic transit gloria mundi.

In that brilliant flight of scientific fancy, the Time Machine, Mr. H.G. Wells has pictured the closing years of the earth in some such long-drawn agony as this. He has given us a vision of a desolate beach by a salt and almost motionless sea. Foul monsters of crab-like form crawl slowly about, beneath a huge hull of sun, red and fixed in the sky. The rocks around are partly coated with an intensely green vegetation, like the lichen in caves, or the plants which grow in a perpetual twilight. And the air is now of an exceeding thinness.

He dips still further into the future, and thus predicts the final form of life:--

"I saw again the moving thing upon the shoal--there was no mistake now that it was a moving thing--against the red water of the sea. It was a round thing, the size of a football perhaps, or it may be bigger, and tentacles trailed down from it; it seemed black against the weltering blood-red water, and it was hopping fitfully about."

What a description of the "Heir of all the Ages!"

To picture the end of our world as the result of a cataclysm of some kind, is, on the other hand, a form of speculation as intensely dramatic as that with which we have just been dealing is unutterably sad.

It is not so many years ago, for instance, that men feared a sudden catastrophe from the possible collision of a comet with our earth. The unreasoning terror with which the ancients were wont to regard these mysterious visitants to our skies had, indeed, been replaced by an apprehension of quite another kind. For instance, as we have seen, the announcement in 1832 that Biela's Comet, then visible, would cut through the orbit of the earth on a certain date threw many persons into a veritable panic. They did not stop to find out the real facts of the case, namely, that, at the time mentioned, the earth would be nearly a month's journey from the point indicated!

It is, indeed, very difficult to say what form of damage the earth would suffer from such a collision. In 1861 it passed, as we have seen, through the tail of the comet without any noticeable result. But the head of a comet, on the other hand, may, for aught we know, contain within it elements of peril for us. A collision with this part might, for instance, result in a violent bombardment of meteors. But these meteors could not be bodies of any great size, for the masses of comets are so very minute that one can hardly suppose them to contain any large or dense constituent portions.

The danger, however, from a comet's head might after all be a danger to our atmosphere. It might precipitate, into the air, gases which would asphyxiate us or cause a general conflagration. It is scarcely necessary to point out that dire results would follow upon any interference with the balance of our atmosphere. For instance, the well-known French astronomer, M. Camille Flammarion,[39] has imagined the absorption of the nitrogen of the air in this way; and has gone on to picture men and animals reduced to breathing only oxygen, first becoming excited, then mad, and finally ending in a perfect saturnalia of delirium.

Lastly, though we have no proof that stars eventually become dark and cold, for human time has so far been all too short to give us even the smallest evidence as to whether heat and light are diminishing in our own sun, yet it seems natural to suppose that such bodies must at last cease their functions, like everything else which we know of. We may, therefore, reasonably presume that there are dark bodies scattered in the depths of space. We have, indeed, a suspicion of at least one, though perhaps it partakes rather of a planetary nature, namely, that "dark" body which continually eclipses Algol, and so causes the temporary diminution of its light. As the sun rushes towards the constellation of Lyra such an extinguished sun may chance to find itself in his path; just as a derelict hulk may loom up out of the darkness right beneath the bows of a vessel sailing the great ocean.

Unfortunately a collision between the sun and a body of this kind could not occur with such merciful suddenness. A tedious warning of its approach would be given from that region of the heavens whither our system is known to be tending. As the dark object would become visible only when sufficiently near our sun to be in some degree illuminated by his rays, it might run the chance at first of being mistaken for a new planet. If such a body were as large, for instance, as our own sun, it should, according to Mr. Gore's calculations, reveal itself to the telescope some fifteen years before the great catastrophe. Steadily its disc would appear to enlarge, so that, about nine years after its discovery, it would become visible to the naked eye. At length the doomed inhabitants of the earth, paralysed with terror, would see their relentless enemy shining like a second moon in the northern skies. Rapidly increasing in apparent size, as the gravitational attractions of the solar orb and of itself interacted more powerfully with diminishing distance, it would at last draw quickly in towards the sun and disappear in the glare.

It is impossible for us to conceive anything more terrible than these closing days, for no menace of catastrophe which we can picture could bear within it such a certainty of fulfilment. It appears, therefore, useless to speculate on the probable actions of men in their now terrestrial prison. Hope, which so far had buoyed them up in the direst calamities, would here have no place. Humanity, in the fulness of its strength, would await a wholesale execution from which there could be no chance at all of a reprieve. Observations of the approaching body would have enabled astronomers to calculate its path with great exactness, and to predict the instant and character of the impact. Eight minutes after the moment allotted for the collision the resulting tide of flame would surge across the earth's orbit, and our globe would quickly pass away in vapour.

And what then?

A nebula, no doubt; and after untold ages the formation possibly from it of a new system, rising phoenix-like from the vast crematorium and filling the place of the old one. A new central sun, perhaps, with its attendant retinue of planets and satellites. And teeming life, perchance, appearing once more in the fulness of time, when temperature in one or other of these bodies had fallen within certain limits, and other predisposing conditions had supervened.

"The world's great age begins anew, The golden years return, The earth doth like a snake renew Her winter weeds outworn: Heaven smiles, and faiths and empires gleam Like wrecks of a dissolving dream.

A brighter Hellas rears its mountains From waves serener far; A new Peneus rolls his fountains Against the morning star; Where fairer Tempes bloom, there sleep Young Cyclads on a sunnier deep.

A loftier Argo cleaves the main, Fraught with a later prize; Another Orpheus sings again, And loves, and weeps, and dies; A new Ulysses leaves once more Calypso for his native shore.

* * * * *

Oh cease! must hate and death return? Cease! must men kill and die? Cease! drain not to its dregs the urn Of bitter prophecy! The world is weary of the past,-- Oh might it die or rest at last!"

[39] See his work, La Fin du Monde, wherein the various ways by which our world may come to an end are dealt with at length, and in a profoundly interesting manner.